T5-AFF-644

ONE WORLD *or* NONE

These Are the Contributors

to

ONE WORLD *or NONE*

H. H. ARNOLD

HANS BETHE

NIELS BOHR

ARTHUR H. COMPTON

E. U. CONDON

ALBERT EINSTEIN

IRVING LANGMUIR

WALTER LIPPMANN

PHILIP MORRISON

J. R. OPPENHEIMER

LOUIS RIDENOUR

FREDERICK SEITZ

HARLOW SHAPLEY

LEO SZILARD

HAROLD C. UREY

EUGENE P. WIGNER

GALE YOUNG

ONE WORLD or NONE

Edited by Dexter Masters *and*
Katharine Way

Contributors

Foreword by Niels Bohr

Introduction by Arthur H. Compton

H. H. Arnold
Hans Bethe
E. U. Condon
Albert Einstein
Irving Langmuir
Walter Lippmann
Philip Morrison
J. R. Oppenheimer
Louis Ridenour
Frederick Seitz
Harlow Shapley
Leo Szilard
Harold Urey
Eugene P. Wigner
Gale Young

and the Federation of American (Atomic) Scientists

Essay Index Reprint Series

Originally Published by
McGRAW-HILL BOOK CO., INC.

BOOKS FOR LIBRARIES PRESS
FREEPORT, NEW YORK

Library of Congress Cataloging in Publication Data

Masters, Dexter, ed.
One world or none.

(Essay index reprint series)
1. Atomic bomb. 2. Atomic energy. I. Way, Katharine, 1903- joint ed. II. Title.
[UF767.M3 1972] 355.8'2 71-37858
ISBN 0-8369-2610-2

PRINTED IN THE UNITED STATES OF AMERICA
BY
NEW WORLD BOOK MANUFACTURING CO., INC.
HALLANDALE, FLORIDA 33009

ARTHUR H. COMPTON, *Chancellor of Washington University, St. Louis, was awarded the Nobel Prize in 1927 for cosmic-ray research. He directed the Metallurgical Laboratory at Chicago where some of the most important discoveries relating to the atomic project were made.*

Introduction

by ARTHUR H. COMPTON

IT was inevitable that mankind should have atomic fire. The worldwide growth of science and technology is the main line of the rapid evolution of man into a social being whose community is the world. The release of atomic energy is but a dramatic step in this evolution. It is a part of our age-old quest to use the forces of nature for shaping the world according to our desire.

No group of men had the power to prevent the coming of the atomic age. The only choice was whether these new powers should first be placed in the hands of the nations that were fighting to preserve their freedom, or should be used by some other group to arm itself with atomic might. It was feared that the other group might be an enemy whose object would have been to enslave the world. The intense incentive of self-preservation was accordingly responsible for making available atomic energy perhaps a decade or two earlier than it might otherwise have come. Thus it was that the Promethean gift was first presented to nations that are conscious of their responsibilities to mankind for its wise use.

The terrific blast at Hiroshima shocked the world into a realization that catastrophe lies ahead if war is not eliminated. This great fear has for the time being overshadowed the hope that atomic energy may vastly enrich human life if given a chance. We now have before us the clear choice between adjusting the pattern of our society on a world basis so that wars cannot come again, or of following the outworn tradition of national self defense, which if carried through to its logical conclusion must result in catastrophic conflict.

Though in broad outline the choice is clear, the manner in which the outline is to be filled in depends upon many obscure factors. These include a knowledge of what the true possibilities of atomic energy are. These possibilities must be known before we can pass intelligent legislation or plan for new industries. In a military sense, what can atomic energy do? In its application to the everyday problems of human life, what promise does it hold? What are the practical possibilities of international agreement that can make us safe in a world of atoms?

It is to give help in answering such problems that this volume is presented. The writers are persons who have been actively concerned with problems of the atomic nucleus, some of them over many years. The technical aspects of the problems are presented in understandable as well as authoritative form by men who have themselves been responsible for turning atomic energy to practical use. They are leaders in atomic engineering. Those who describe the military effectiveness of atomic energy have followed the development of the bomb from the beginning and have had a first-hand view of its effects. Those who discuss its political implications are men who have held this problem to their hearts for years.

The suggestions put forward with regard to the national and international control of atomic energy are the views of the individuals who propose them and do not necessarily represent the views of all of the contributors to this book. The opinions are nevertheless the results of mature thinking by well-informed individuals. It is hoped that their presentation will help us to understand the issues involved and why certain sacrifices such as that of national sovereignty are called for.

It is doubtful whether the world has faced a more critical problem than that of the proper handling of atomic energy. Our great hope is that this volume will help in finding a wise solution to that problem, a solution which will bring us lasting peace and make real to us the clear possibility of an enriched human life.

CONTENTS

Foreword

NIELS BOHR, *whose nuclear research did much to create the atomic age and won him the Nobel Prize in 1922 at the age of 37, escaped from the Nazis in his native Denmark in 1943. Coming to America, he played a major part in the development of the uranium project. He has now returned to Denmark.*

Science and Civilization*

by NIELS BOHR

THE possibility of releasing vast amounts of energy through atomic disintegration, which means a veritable revolution of human resources, cannot but raise in the mind of everyone the question of where the advance of physical science is leading civilization. While the increasing mastery of the forces of nature has contributed so prolifically to human welfare and holds out even greater promises, it is evident that the formidable power of destruction that has come within reach of man may become a mortal menace unless human society can adjust itself to the exigencies of the situation. Civilization is presented with a challenge more serious perhaps than ever before, and the fate of humanity will depend on its ability to unite in averting common dangers and jointly to reap the benefit from the immense opportunities which the progress of science offers.

In its origin science is inseparable from the collecting and ordering of experience, gained in the struggle for existence, which enabled our ancestors to raise mankind to its present position among the other living beings that inhabit our earth. Even in highly organized communities where, within the distribution of labor, scientific study has become an occupation by itself, the progress of science and the advance of civilization have remained most intimately interwoven. Of course, practical needs are still an impetus to scientific research, but it need hardly be stressed how often technical developments of the greatest importance for civilization have originated from studies aimed only at augmenting our knowledge and deepening our understanding. Such endeavors know no national borders, and where one scientist has left the trail another has taken it up, often in a distant part of the world. For scientists have long considered themselves a brotherhood working in the service of common human ideals.

In no domain of science have these lessons received stronger emphasis than in the exploration of the atom, which just now is bearing consequences of such overwhelming practical implications. As is well known, the roots of the idea of atoms as the ultimate constituents of matter go back to ancient thinkers searching for a foundation to explain the regularity which, in spite of all variability, is ever more clearly revealed by the study of natural phenomena. After the Renaissance, when science entered so fertile a period, atomic theory gradually became of the greatest importance for the physical and chemical sciences, although until half a century ago it was generally accepted that, owing to the coarseness of our senses, any direct proof of the existence of atoms would always remain beyond human scope. Aided, however, by the refined tools of modern technique, the development of the art of experimentation has removed such limitation and even yielded detailed information about the interior structure of atoms.

In particular, the discovery that almost the entire mass of the atom is concentrated in a central nucleus proved to have the most far-reaching consequences. Not only did it become evident that the remarkable stability of the chemical elements is due to the immutability of the atomic nucleus when exposed to ordinary physical agencies, but a novel field of research was opened up by the study of the special conditions under which distintegrations of the nuclei themselves may be brought about. Such processes, whereby the very elements are transformed, were found to differ fundamentally in character and violence from chemical reactions, and their investigation led to a rapid succession of important discoveries through which

* This statement appeared in *The London Times,* August 11, 1945. It is published here, for the first time in this country, by special arrangement with Professor Bohr.

ultimately the possibility of a large-scale release of atomic energy came into sight. This progress was achieved in the course of a few decades and was due not least to most effective international cooperation. The world community of physicists was, so to speak, welded into a single team, rendering it more difficult than ever to disentangle the contributions of individual workers.

The grim realities being revealed to the world these days will no doubt, in the minds of many, revive the terrifying prospects forecast in fiction. With all due admiration for such imagination, it is, however, most essential to appreciate the contrast between these fantasies and the actual situation confronting us. Far from offering any easy means to bring destruction forth, as it were by witchcraft, scientific insight has made it evident that use of nuclear disintegration for devastating explosions demands most elaborate preparations, involving a profound change in the atomic composition of the materials found on earth. The astounding achievement of producing an enormous display of power on the basis of experience gained by the study of minute effects, perceptible only by the most delicate instruments, has in fact, besides a most intensive research effort, required an immense engineering enterprise, strikingly illuminating the potentialities of modern industrial development.

Indeed, not only have we left the time far behind where each man, for self-protection, could pick up the nearest stone, but we have even reached the stage where the degree of security offered to the citizens of a nation by collective defense measures is entirely insufficient. Against the new destructive powers no defense may be possible, and the issue centers on world-wide cooperation to prevent any use of the new sources of energy that does not serve mankind as a whole. The possibility of international regulation for this purpose should be ensured by the very magnitude and the peculiar character of the efforts that will be indispensable for the production of the formidable new weapon. It is obvious, however, that no control can be effective without free access to full scientific information and the granting of the opportunity of international supervision of all undertakings that, unless regulated, might become a source of disaster.

Such measures will, of course, demand the abolition of barriers hitherto considered necessary to safeguard national interests but now standing in the way of common security against unprecedented dangers. Certainly the handling of the precarious situation will demand the good will of all nations, but it must be recognized that we are dealing with what is potentially a deadly challenge to civilization itself. A better background for meeting such a situation could hardly be imagined than the earnest desire to seek a firm foundation for world security, so unanimously expressed by all those nations which only through united efforts have been able to defend elementary human rights. The extent of the contribution that agreement on this vital matter would make to the removal of obstacles to mutual confidence, and to the promotion of a harmonious relationship between nations can hardly be exaggerated.

In the great task lying ahead, which places on our generation the gravest responsibility toward posterity, scientists all over the world may offer most valuable services. Not only do the bonds created through scientific intercourse form some of the firmest ties between individuals from different nations, but the whole scientific community will surely join in a vigorous effort to induce in wider circles an adequate appreciation of what is at stake and to appeal to humanity at large to heed the warning that has been sounded. It need not be added that every scientist who has taken part in laying the foundation for the new development or who has been called upon to participate in work that might have proved decisive in the struggle to preserve a state of civilization where human culture can freely develop is prepared to assist, in any way open to him, in bringing about an outcome of the present crisis of humanity that is worthy of the ideals for which science through the ages has stood.

ONE WORLD *or* NONE

Chapter 1

Philip Morrison, *now professor of physics at Cornell University, was active on the atomic bomb project at Chicago and Los Alamos. At the request of the War Department, he went to Japan to investigate the effects of the Hiroshima bomb.*

If the Bomb Gets Out of Hand

by PHILIP MORRISON

WE SAT in a small open wooden hut, like a booth at a church fair, listening to the Japanese General Staff major from Tokyo. Around us the ground was blackened. The trees were strangely bare for September beside the Inland Sea. The advance party of the American Army mission to study the effects of the atom bomb had come to Hiroshima. In the rubble of the castle grounds, the old headquarters of the Fifth Division, the local authorities had prepared for us a meeting with the men who had lived through the disaster of the first atomic bomb. The major was very young and very grave. He spoke slowly and carefully, like a man who wants to be properly translated and clearly understood. The story he told is worth hearing. It is the story of the first impact of the atomic bomb on the structure of a nation.

About a quarter-past seven on Monday morning, August 6, the Japanese early-warning radar net had detected the approach of some enemy aircraft headed for the southern part of Honshu, and doubtless for the ports of the Inland Sea. The alert was given, and radio broadcasting stopped in many cities, among them Hiroshima. The raiders approached the coast at very high altitude. At nearly eight o'clock the radar operators determined that the number of planes coming in was very small—probably not more than three—and the air raid alert was lifted. The normal broadcast warning was given to the population that it might be advisable to go to shelter if B-29's were actually sighted, but that no raid was expected beyond some sort of reconnaissance. At 8:16 the Tokyo control operator of the Japan Broadcasting Corporation noticed that the Hiroshima station had gone off the air. He tried to use another telephone line to re-establish his program, but it too had failed. About twenty minutes later the Tokyo railroad telegraph center realized that the main line telegraph had stopped working just north of Hiroshima. And from some small railway stops within ten miles of that city there had come unofficial and rather confused reports of a terrible explosion in Hiroshima. All these events were then reported to the air-raid defense headquarters of the General Staff. The military called again and again the Army wireless station at the castle in Hiroshima. There was no answer. Something had happened in Hiroshima. The men at headquarters were puzzled. They knew that no large enemy raid could have occurred; they knew that no sizeable store of explosives was in Hiroshima at that time.

The young major of the General Staff was ordered in. He was instructed to fly immediately by army plane to Hiroshima, to land, to survey the damage, and to return to Tokyo with reliable information for the staff. It was generally felt in the air-raid defense headquarters that nothing serious had taken place, that the nervous days of August, 1945, in Japan had fanned up a terrible rumor from a few sparks of truth. The major went to the airport and took off for the southwest. After flying for about three hours, still nearly one hundred miles from Hiroshima, he and his pilot saw a great cloud of smoke from the south. In the bright afternoon Hiroshima was burning. The major's plane reached the city. They circled in disbelief. A great scar, still burning, was all that was left of the center of a busy city. They flew over the military landing strip to land, but the installations below them were smashed. The field was deserted.

About thirty miles south of the wrecked city is the large naval base of Kure, already battered by carrier strikes from the American fleet. The major landed at the Kure airfield. He was welcomed by the naval officers there as the first official representative of aid from Tokyo. They had seen the explosion at Hiroshima. Truckloads of sailors had been sent up to help the city in this strange disaster, but terrible fires had blocked the roads, and the men had turned back. A few refugees had straggled out of the northern part of the town, their clothes and skin burned, to tell near-hysterical stories of incredible violence. Great winds blew in the streets, they said. Debris and the dead were everywhere. The great explosion had been for each survivor a bomb hitting directly on his house. The staff major, thrown into the grimmest of responsibilities, organized some two thousand sailors into parties, which reached the city about dusk. They were the first group of rescue workers to enter Hiroshima.

The major took charge for several days. The rail line was repaired, and trainloads of survivors were shipped north. The trains came first from Onomichi, where, about forty miles north, there was a large naval hospital. Soon the hospital was filled, and its movable supplies exhausted. Then the trains bore the injured still farther north, until there too the medical facilities were completely used up. Some sufferers were shipped twenty-four hours by train before they came to a place where they might be treated. Hospital units were mobilized by Tokyo to come from hundreds of miles to set up dressing stations in Hiroshima. One bomb and one plane had reduced a city of four hundred thousand inhabitants to a singular position in the war economy of Japan: Hiroshima consumed bandages and doctors, while it produced only trainloads of the burned and the broken. Its story brought terror to all the cities of the islands.

The experts in the science of the killing of cities have developed a concept which well describes the disaster of Hiroshima, the disaster which will come to any city which feels the atomic bomb. That is the idea of saturation. Its meaning is simple: if you strike at a man or a city, your victim defends himself. He hits you, he throws up flak, he fights the fires, he cares for the wounded, he rebuilds the houses, he throws tarpaulins over the shelterless machinery. The harder you strike, the greater his efforts to defend himself. But if you strike all at once with overwhelming force, he cannot defend himself. He is stunned. The city's flak batteries are all shooting as fast as they can; the firemen are all at work on the flames of their homes. Then your strike may grow larger with impunity. He is doing his utmost, he can no longer respond to greater damage by greater effort in defense. The defenses are saturated.

The atomic bomb is pre-eminently the weapon of saturation. It destroys so large an area so completely and so suddenly that the defense is overwhelmed. In Hiroshima there were thirty-three modern fire stations; twenty-seven were made useless by the bombing. Three-quarters of the fire-fighting personnel were killed or severely injured. At the same instant, hundreds, perhaps thousands, of fires broke out in the wrecked area. How could these fires be brought under control? There were some quarter of a million people injured in a single minute. The medical officer in charge of the public health organization was buried under his house. His assistant was killed, and so was *his* assistant. The commanding officer of the military was killed, and his aide, and his aide's aide, and in fact every member of his staff. Of 298 registered physicians, only thirty were able to care for the survivors. Of nearly twenty-four hundred nurses and orderlies, only six hundred were ready for work after the blast. How could the injured be treated or evacuation properly organized? The power substation which served the center of the city was destroyed, the railroad was cut, and the rail station smashed and burned. The telephone and telegraph exchange was wrecked. Every hospital but one in the city was badly damaged; not one was able to shelter its patients from the rain—even if its shell of concrete still stood—without roof, partitions, or casements. There were whole sections of the outer city undamaged, but the people there were unable to give effective aid, lacking leadership, organization, supplies, and shelter. The Japanese defenses had already been proved inadequate under the terrible fire raids of the B-29's, which had desolated so many of Japan's cities. But under the atomic bomb their strained defenses came to complete saturation. At Nagasaki, the target of the second atomic bomb, the organization of relief was even poorer. The people had given up.

A Hiroshima official waved his hand over his wrecked city and said: "All this from one bomb; it is unendurable." We knew what he meant. Week after week the great flights of B-29's from the Marianas had laid flame to the cities of all Japan. But at least there was a warning. You knew when the government announced a great raid in progress that, though Osaka people would face an infernal night, you in Nagoya could sleep. For the raids of a thousand bombers could not be hidden, and the fire raid had formed a pattern. But every day over any city of the chain there was a chance for a few American planes to come. These inquiring planes had been photographers or weather forecasters or even occasionally nuisance raiders; never before had a single plane destroyed a city. Now, all this was changed. From any plane casually flying almost beyond the range of flak there could come death and flame for an entire city. The alert would have to be sounded now night and day in every city. If the raiders were over Sapporo, the people of Shimonoseki, a thousand miles away, must still fear even one airplane. This is unendurable.

If war comes again, atomic war, there will not even be the chance for alerts. A single bomb can saturate a city the size of Indianapolis, or a whole district of a great city, like Lower Manhattan, or Telegraph Hill and the Marina, or Hyde Park and the South Shore. The bombs can come by plane or rocket in thousands, and all at once. What measures of defense can there be? To destroy the bombs in flight many measures will be attempted, but they cannot be a hundred per cent effective. It is not easy to picture what even one single bomb will do. We saw the test shot in the New Mexico desert, and we pored over and calculated the damage that a city would suffer. But on the ground at Hiroshima and Nagasaki there lies the first convincing evidence of the damage done by the present atomic bomb.

The streets and the buildings of Hiroshima are unfamiliar to Americans. Even from pictures of the damage realization is abstract and remote. A clearer and truer understanding can be gained from thinking of the bomb as falling on a city, among buildings and people, which Americans know well. The diversity of awful experience which I saw at Hiroshima, and which I was told about by its citizens, I shall project on an American target. Please do not believe that there is exaggeration here; this story will be conservative, it will allow for no increase in the effectiveness of the bomb. It will tell of only one where, if there is atomic war, twenty will fall. Your city, too, is a good target.

The microwave early-warning radar towers on the Jersey coast and up past Riverside had recorded the approach of the missile. It was 12:07 when they noted the end of the signal, and the operators wondered what the thing had been. When the telephone circuits failed and the teletype stopped, they grew worried. When they listened to the shaky and disturbed news report from WABC a few minutes later, they knew what had made the mark on the screen. One of the men walked outside with his camera and looked north in the bright noon sun to see the great pillar of cloud he knew would come. The wind had been from the northwest all day, and it is interesting to note that the radioactive cloud passed over the same radar installation which had first remarked the missile. The recording radiation meter at the station showed a harmless quantity of gamma radiation, but the photographic film was badly fogged.

The device detonated about half a mile in the air, just above the corner of Third Avenue and East 20th Street, near Gramercy Park. Evidently there had been no special target chosen, just Manhattan and its people. The flash startled every New Yorker out of doors from Coney Island to Van Cortlandt Park, and in the minute it took the sound to travel over the whole great city, millions understood dimly what had happened.

The district near the center of the explosion was incredible. From the river west to Seventh Avenue, and from south of Union Square to the middle thirties, the streets were filled with the dead and dying. The old men sitting on the park benches in the square never knew what had happened. They were chiefly charred black on the side toward the bomb. Everywhere in this whole district were men with burning clothing, women with terrible red and blackened burns, and dead children caught while hurrying home to lunch. The thousands of brick and brownstone walk-ups, huddled closely to the elevated and packed thickly between the rivers, were badly shaken in a few seconds. The parapets and the porches tumbled into the streets, the glass of the windows blew sometimes out and sometimes in, depending on the complex geometry of the old buildings. The plaster fell on the heads of the tenants, old floors and stairs collapsed under the terrible wind of the blast, and only the heavy walls stood to mark the homes. Closer to the center nothing much was left. Many of the narrow streets passing between the old five-floor brick or stone tenements were choked with rubble, until it was difficult to walk down the street. Here and there collapsed buildings had piled a great heap of pitiful debris, all the wares and effects of living, into a useless and smoldering jumble. Everywhere there were fires, usually licking at already useless wreckage, but making heartbreakingly difficult the escape of the injured and the slow work of the half-stunned rescue parties.

The elevated structure stood up comparatively well. All the elevated stations from Fourteenth almost to midtown were wrecks. The steps were gone, the flimsy flooring and the old baroque railings lay in the street below. Only the clean steel frames were for the most part intact. In the blocks near Twenty-third even the main frame had gone, and the twisted vertical columns remained above the nightmare of steel below. The loss of life was very large from this alone. A train had been pushed off going at full speed north on Second Avenue near Twentieth, and the flames which burned the whole of the district seemed to begin from the wreckage. A few concrete garages and warehouses stood up over the gaunt frames of the elevated tracks, but the whirlwind which tore through them left the interiors wrecked. Fire usually finished the job.

The great buildings were not destroyed; none had been very close to the blast. But they were not unharmed. The high Metropolitan Tower was the worst damaged. The steelwork stood unharmed nearly to the top, though it was badly twisted where a whole ten-story wall section had come down into the street. The interior partitions from the sixteenth floor and up were completely gone, and even some floors had failed, leaving a kind of half-filled honeycomb of a building above the twentieth floor. More than fifty people were later said to have managed to clamber down from the wreck. It is known that eighteen of the radiation deaths recorded in the St. Louis hospitals later were of people who had been in the higher floors of this building when the bomb struck. The people below the tenth floor were not fatally injured for the most part. Fractures and lacerations from glass were the principal cause of injury. A good many hundreds of people from the south side of the building died two or three weeks after the blast from radiation. Among them was a well-known aeronautical engineer who had managed to remain uninjured by the flash burn or the blast, standing as he was behind a steel beam column on the south side of the first floor, near the windows. He bravely worked the whole day as one of the rescue party for the Tower. The bad nauseous symptoms which he underwent at six o'clock caused him to seek hospitalization at Philadelphia, where he died in twelve days, while working on a report for the Air Forces on the extent of the damage to steel structures.

The Empire State building nearly a mile away was strikingly little damaged. The radio structures and the external ornament of the high spire were swept clean. The windows of course were shattered and much damage was done to the light partitions and even to the glassy exterior walls on the higher floors. Elevator machinery was badly damaged by a freakish falling beam and many were trapped in immobile cars. The flash scorched papers and window screens and set fires going in all the offices on the side facing the blast. These fires were brought under control in a day or so. For months after the blast the high tower seemed to stand defiantly at the upper edge of the vanished district, but the building was useless except in the very lowest floors. The tenants of the building had not fared so well as its steel and concrete frame; the great dressing station established in the corridors and rooms of the first five floors handled many of them, and sent many of them to the Police Department's common graves.

The underground world of the city had been relatively safe. When the power failed in the whole lower eastern Manhattan district because of the destruction of the transformer sub-stations, the subway power alone was restorable. The Lexington gratings collapsed, and near the blast one or two large street cave-ins had stopped traffic on the IRT and flooded part of the tubes from broken mains. But the greater number of subway passengers and crews escaped. A few hundred were trampled in a bad panic at the Thirty-fourth Street entrance, and one train piled into the wreckage below ground very near the aiming point. Some people walked north underground all the way to the Bronx, not believing it safe to come up any closer to the bomb. Men in the sub-basements of the great buildings were horrified when they came up to see why the lights had failed; they had known nothing of the great blast but a ground tremor and the dust of falling plaster.

The nearness of Bellevue Hospital to the blast

—about half a mile—was tragic. The long brick walls collapsed. Only a few patients here and there survived. The doctors and nurses had no time to salvage even the carefully prepared emergency supplies. Fire attacked the ruin, and the scenes which followed are indescribable. The knocking out of Bellevue was a hard blow to the rescue organization of the city and delayed for some time the proper organization of relief.

There were many stories of unbelievable good fortune and magnificent heroism. One man, a glassblower apprentice, was walking along Lexington south to Twenty-fourth. He described the great flash, but he was protected from a direct view by the corner of a building. The blast knocked him down along the broad street, but no heavy object hit him, and he escaped without serious injury. All day and night he worked, leading the badly injured north and pulling many people from the wreckage. Though he was only a few hundred yards from the point below the point of impact, he suffered no symptoms of radiation injury. He was the only person on the streets of the city within a ten-block radius who is known to have survived without serious injury, and not more than a thousand of the hospitalized but recovering victims were as close as he.

The most tragic of all the stories of the disaster is that of the radiation casualties. They included people from as far away as the Public Library or the neighborhood of Police Headquarters downtown, but most of them came from the streets between the river and Fifth Avenue, from Tenth or Twelfth to the early Thirties. They were all lucky people. Most of them had had remarkable escapes from fire, from flash burns, from falling buildings. The people around them had never gotten away, but they had crawled, injured but alive, from the wreckage of homes or shops, from the elevated platforms, or from cellar stairways. Some had seen the great flash, felt the floor collapse, and picked themselves up ten minutes later from the rubble of their homes. Others had gotten free of the bus or auto they were in when it was thrown into a wall and had pulled out after them the dead and dying who had been their fellow passengers. They were all lucky, as they said. Some were dramatically uninjured, like the aeronautical engineer. But they all died. They died in the hospitals of Philadelphia, Pittsburgh, Rochester, and St. Louis in the three weeks following the bombing. They died of unstoppable internal hemorrhages, of wildfire infections, of slow oozing of the blood into the flesh. Nothing seemed to help them much, and the end was neither slow nor very fast, but sure. They were relatively few in number—the doctors quarreled for months about their census, but it was certainly twenty thousand, and it may have been many more.

The people far away who lived through the days of the aftermath suffered too. Homes and offices were systematically and badly damaged as far away as Fifty-seventh Street and Fulton Market, and across both rivers. Every block had its collapsed brick structures and its many walls carried away, and its dozen dead. There were not many windows intact on Manhattan Island, and there were many thousands wearing the face dressings that marked the target of glass splinters. But their lives went on, the damage was slowly repaired, and those who had no job to do there stayed far away from the scar that had been the Twenties. The rerouting of traffic and the repair of the telephone and the electrical and the water systems had its effect on the economic life of the whole city. The damage was felt in many ways as a drain on the recuperative power of the whole of New York City, and the loss of one-tenth of the people and the property of the city was enough to lessen the work of the city by half. People moved away and tried to forget.

The statistics were never very accurate. About three hundred thousand were killed, all agreed. At least two hundred thousand had been buried and cremated by the crews of volunteer police and of the Army division sent in. The others were still in the ruins, or burned to vapor and ash. As many again were seriously injured. They clogged the hospitals of the East and turned many a Long Island and New Jersey resort that summer into a hospital town.

There was no one of the eight million who had not his story to tell. The man who saw the blast through the netting of the monkey cage in Central Park, and bore for days on the unnatural ruddy tan of his face the white imprint of the shadow of the netting, was famous. The amateurs who collected radioactive souvenirs from the strong patch of radioactivity which sickened Greenwich Villagers for weeks were matched by those who found scorched shadow patterns in the wallpaper and plasterboard of a thousand wrecked homes.

New York City had thus suffered under one bomb, and the story is unreal in only one way: The bombs will never again, as in Japan, come in ones or twos. They will come in hundreds, even in thousands. Even if, by means as yet unknown, we are able to stop as many as 90 per cent of these missiles, their number will still be large. If the bomb gets out of hand, if we do not learn to live together so that science will be our help and not our hurt, there is only one sure future. The cities of men on earth will perish.

Chapter 2

HARLOW SHAPLEY, *one of our most distinguished astronomers, is director of the Harvard College Observatory. Dr. Shapley is the author of numerous books and articles on astronomy, and here he tells the story of atomic transmutation in the stars.*

It's an Old Story with the Stars

by HARLOW SHAPLEY

THE historians of the next millennium should be able to chronicle correctly the fact that atomic energy came into man's civilization and into his practical economy in the first half of the twentieth century. But there must be less precision in our recording that the production of atomic energy entered the economy of the stars in the year 3,000,000,000 B. C. Although there is uncertainty in the date, there is no doubt of the fact. The release and use of atomic energy is an old story with the sun and other stars; it was accepted as a fundamental tenet of astrophysics two or three decades before the splitting of the uranium atom was accomplished willfully on one of the sun's planets.

The story of atomic transmutation in the stars is worth sketching in a volume on the origins, nature, and responsibilities of the new atomic age. It is worth recounting for two excellent reasons. One is that atomic transmutation and energy release in the stars are phenomena at the very core, pith, and root of fundamental knowledge of the universe. An acceptable cosmogony must be based on the interrelation of matter and radiation, and it must satisfactorily account for these relationships throughout all space and all time.

The second attraction of this story comes from the curious fact that it was the fossil bones and plant leaves in our terrestrial rocks that led us to knowledge of the atom splitting in the stars.

This is how it came about. The astronomers long ago began to measure the total output of light from the sun. They knew of the great distance separating the solar surface and the earth's surface—more than 92,000,000 miles. They could measure directly how much radiant energy comes in each second to each square mile of the earth's surface. The measurements and calculations are simple. As seen from the sun, the earth is inconspicuous, exceedingly so; we find, in fact, that it blocks less than one part in two billion of the outpouring solar radiation. Since so much heat and light get to us notwithstanding our minute apportionment, it is obvious that the sun is a stupendous producer. The rate of radiation is so great that thoughtful astronomers of the early nineteenth century became concerned about it. Would the sun exhaust its sources of supply and soon run down? How could this furious energy be kept going long enough to account for the recorded past and provide for the hoped-for future?

The primitive supposition that the sun simply is burning up in the fashion that coal oxidizes in our stoves never attained scientific standing. The better suggestion was advanced that meteors and comets falling into the sun might provide an adequate energy supply through friction and impact in the solar atmosphere. Later Helmholtz and his followers pointed to a sufficient source of solar energy in the natural and automatically regulated contraction of the gaseous sun. Isolated in cold space, the sun tends to cool off from radiating away its heat. The cooling results in a tendency to shrink, with the atmosphere's condensing toward the sun's center. But the shrinking transforms energy of position into energy for radiation, and the sun in consequence tends to warm up. The combination of processes results in maintaining the constancy of the outpouring radiation.

But it turns out that there is something wrong with this solution. Here is where the fossils enter and also the phenomena of radioactivity. Here, in fact, is where the renowned element uranium first enters the picture of atomic energy and the fuel economy of the stars.

Geologists have long been cautious souls. They find fossils of animals and plants in many rocks in many lands. Evidences of great age are abundant. But how old are these deep-lying beds that contain vestiges of extinct animals and plants? Considerably bedeviled by the theologians of the nineteenth century, geologists were long content to estimate paleozoic ages in the tens and hundreds of thousands of years; later, in a modest number of millions of years.

Then Becquerel and the Curies came upon the scene, and with them soon came radium and the sensational property of natural radioactivity in radium, thorium, actinium, and uranium. The *artificial splitting* of the uranium atom, which now concerns us deeply, was far in the future. But the *natural chipping* of the uranium atom, its automatic transmutation into a somewhat lighter element through the expulsion of an alpha particle (nucleus of the helium atom), was already recognized in 1896. And it was soon deduced that this chipping had been going on throughout all the time that the earth has existed and perhaps longer. The nucleus of the atom revealed its internal powers fifty years ago when its emission of the intense radiation and high-speed particles that come as a by-product of natural radioactivity was first noted.

When we first treated tumors with radium, we first used the power of the atomic nucleus. Then it was that the children of the earth first followed the stars of the heavens in living with atomic energy in the raw. They stepped on the threshold of the atomic age to which the door has now been thrown wide open.

The particular use of uranium that is of most interest to this part of our story lies in its utility as a clock for measuring the age of rocks. Its continuous and inevitable decay, toward the end products of lead and helium, makes uranium especially important as a measure of the time that has gone. The more helium and the more lead one finds in a uranium-bearing rock, the longer this uranium has been operating and the older the rock.

The first measurements produced revolutionary results. The old caution of the geologists and the paleontologists was dissipated by the new geochemistry. The revised time scale was found to be comfortably long for the slow processes of geological change and biological development. It was consistent with the evidences of erosion and sedimentation. But now the astronomers were put in a hole.

As the sun goes, we say, so go the stars. Solve the problem of solar energy and you have accounted for the radiant operation of the sidereal universe. If contraction works for the sun, it must work elsewhere. So the temporarily satisfactory theory of contraction began to lose standing as a satisfying creed. It was too limited to stand the pressure of time. It was defeated by the archaic ferns of the carboniferous formations, by their newly determined antiquity, and by the geochemical evidence that the ferns and contiguous uranium rocks demand a long, long time scale.

In other words, the sudden realization that the earth's surface is exceedingly old made it necessary to reexamine the theories of the sources of sunlight and especially to ponder the character of ancient sunlight. For it is clear, even to the amateur inspector of paleozoic-life remains, that living conditions (so far as air and light are concerned) were practically the same three hundred million years ago as now and probably also five hundred million years ago. The fossil pickers and the chemists working with radioactivity in the rocks had found unknowingly something greater than they knew.

It has been called serendipity—this faculty or fact of accidentally finding a result of superior significance while searching for something else. It occurs frequently, but seldom has it led to such a magnificent revelation as the exposing of the secret life of the stars. Serendipity operated while the geochemist measured the ratio of lead to uranium in various old strata of rocks and while the paleontologist pondered the structures of fossil plants. For they found results that forced the astrophysicist, in searching vigorously again for a sufficient explanation of constant sunlight, to make a speculative invasion of nuclear physics and to find there a clue to the coming of a terrestrial atomic era. There could be no better example of the intermingling and advisable cooperation of the various highly specialized sciences.

Let us see how the clue was followed by the scientists. In 1904, J. H. Jeans, a far-seeing cosmogonist, suggested that if in their frenzied agitation the electrons and protons of high-temperature matter should collide and exterminate each other, there would be an effective release of energy. (This was twenty-five years before the neutron was accepted as a major constituent of the atomic nucleus and as the equivalent of a proton with its positive charge annihilated.) The next year Einstein and the special theory of relativity provided the wonderworking mathematical machinery that permitted the calculation of just how much energy would be released in the annihilation of matter or in its partial annihilation; for on the new physical principle matter and energy were indeed equivalent. Grams of matter and ergs of energy are the same entity with different make-up.

The Einstein equation states that you can, if you know the trick, get 9×10^{20} ergs per gram—a tremendous output, well illustrated by Hans Bethe's observation that 1 ounce of matter is energetically equivalent to the output for a month of the great power plant at Boulder Dam.

The astronomers decided that the stars do know the trick. In their hot interiors they can transform the atoms of matter into radiant energy which much modifies their surfaces and comes to us as sunlight and starlight. Although the sun in this process must lose by radiation more than four million tons of its mass every second, its total material is so great that it can run steadily for millions of millions of years. The few hundred million years of steady sunshine required by the evidences of paleontology could now be easily provided. In fact, the sun and the stars would not need to annihilate matter completely for their energy supply. They could, for instance, transform only 1 per cent of the mass of their atoms into radiation and amply meet all requirements.

Apparently the stars work on this percentage basis. The transmutation of hydrogen, the lightest element, into helium, the next lightest, with the coincident transformation of a small increment of the mass involved into radiation, is what happens in the hot interiors of stars built on the solar model.

For some of the stars, it seems, the internal temperatures are too low for the hydrogen-helium synthesis to be effective, but other atomic processes are there possible and probable. The other lightweight atoms—deuterium, lithium, beryllium, boron—are now believed to be potential collaborators in the production of energy through atomic transmutation in stellar interiors.

In the sun's interior, where hydrogen builds into helium, the temperature is more than twenty million degrees centigrade. The surface temperature of the sun is only about six thousand degrees centigrade. We protoplasmic organisms can indeed be grateful that the center is not exposed to us. We could not endure such temperatures, which are comparable only with the flash of an atomic bomb. Our existence is possible only because the outer atmospheres of the sun serve as a screen and a radiation softener.

But we should remember that atomic transmutations are at the bottom of terrestrial life. They yield the by-product of comfortable radiation that has been so long enduring that the earth's water, air, and rocks could be kept warm enough for biological evolution to occur. One end result of the atomic processes in the sun is therefore the writing and reading of books on the new atomic age. We have got where we are in the cosmos by the grace of celestial atomic reactions.

Since we are personally so much indebted to the helium synthesis and its by-product of energy for our green leaves and for our warmhearted planet, we ought to mention how the astrochemistry works deep in the sun. It is not a direct process of four high-speed hydrogen atoms merging directly into a helium atom. Another common element, carbon, acts as an intermediary in assembling the hydrogen units.

There are several steps involving several transmutations before the end products of helium and radiation are attained. The ordinary carbon atom first captures a hydrogen nucleus and thereby becomes an isotope of nitrogen. It is heavier than the carbon atom, of course, and is unstable. It decays into a heavy isotope of carbon through a spontaneous radioactive process.* This new carbon atom picks up another hydrogen nucleus and becomes a heavier isotope of nitrogen. The process goes on through additional captures and radioactive changes until the catalyzing carbon atom has annexed four hydrogen nuclei, at which stage it splits back into ordinary carbon and into helium. This "carbon-stove" mechanism has, in a sense, burned hydrogen fuel into helium ash, and at various points in the complicated process it has released atomic energy in the nature of very short-wave radiation. The total mass of the four hydrogen nuclei is greater than is required for one resulting helium nucleus. The excess of matter is about 1 per cent, and it becomes radiation in accordance with Einstein's equivalence principle. It is surplus material that is traded in as energy—at the rate of nine hundred million trillion ergs per gram.

*** For a full account of the synthesis, see Goldberg and Aller, *Atoms, Stars, and Nebulae*, pp. 269 *ff.*, Philadelphia, The Blakiston Company, 1940.**

For the thousands of millions of years that the stars have been shining, the majority have been living on this catalytic activity of carbon that transforms the most common element in the universe, hydrogen, into helium and other atoms that are larger and more complicated. Our current interpretation of the atomic mechanisms within stars will no doubt be refined as further theory and observation become available. The picture here sketched reflects the work over a decade or two of many astrophysicists and atomic physicists, but the major contributor has been Dr. Hans Bethe—a contributor also of a later chapter in this volume.

The stellar atom processes work among the lighter species of atoms, while the human atom-splitters of the past few years have worked among the heaviest of the atoms at the upper end of the ninety-two elements. And they do not stop with No. 92; they now go several steps further. No. 93 is neptunium, named for the planet Neptune, as uranium (No. 92) was named for Uranus, and thorium (No. 91) was named for Jupiter (Thor). The recently discovered plutonium (No. 94) had its name ready at hand—by a rather narrow margin, because the outermost known planet, Pluto, was discovered and named only in 1930. The newly discovered elements 95 and 96 catch the astronomers unprepared. If eventually we do find two other planets in the outer reaches, it is foreordained that the names assigned to them will be the names selected for elements 95 and 96. We are just that sentimental.

The release of atomic energy on a grand scale among the stars is not limited to the orderly processes that have maintained starlight steadily throughout geologic ages. Certain peculiar phenomena of the solar corona may be attributable in part to an atom-splitting process at or near the surface of the sun. And it is generally accepted that the novae, or at least the infrequent supernovae, deserve attention when the atomic-energy problems of the universe are considered. The supernovae may indicate what might happen to one, whether star or man, who plays around carelessly with atomic energy and lets it get out of hand. In stellar interiors the pressures, temperatures, radiation densities, and chemical constitution are all interrelated and all involved in the maintenance of the steady state that characterizes most stars. The production of energy from matter, if a star is a steady performer, demands certain equilibrium conditions. Otherwise something drastic may happen. Stars like the sun seem to manage very well. They are not even afflicted by discernible periodic pulsations and variations in size or in surface temperature as are so many of the giant stars.

It appears to be otherwise, however, with the supernovae. Something inside a star has upset the equilibrium between production and distribution. All the evidence points to violent and disastrous explosion. Without previous indication that something is wrong, the star's surface begins to expand with great velocity, and the surface temperature rises. Within a few hours the brightness increases so rapidly that it frequently excels, at maximum effort, the light of one hundred million stars of the sun's brightness.

As the explosive outburst subsides in the course of hours or days, we find, in the few cases that could be properly studied, wreckage of the violence. The outer part of the star has been blown out into space in all directions. Nebulosity frequently appears. The well-known Crab Nebula in Taurus is now recognized as the wreckage of the supernova of July 4, A.D. 1054. The Orientals of that time recorded an enormously bright "temporary star." It excelled in brightness all the stars in the sky but soon faded away, not to be seen again until the groping telescopes of seven centuries later began to list faint nebulous objects. In the position where the great star of 1054 had transiently glittered was an irregular nebulosity. Modern telescopes show that it is a mass of gases, still expanding—the result, apparently, of the mishandling by a star of its resources in atomic energy.

Chapter 3

Eugene P. Wigner, *professor of physics at Princeton, was one of the group originally responsible for getting government support for the atomic bomb project. He was concerned from the first with the physics of chain-reacting piles and in 1942 moved to Chicago to head the theoretical physics work at the Metallurgical Laboratory.*

Roots of the Atomic Age

by EUGENE P. WIGNER

ONLY very few of us can design or construct a steam engine or prepare an explosive compound, and it is not the purpose of this chapter to serve as a textbook on atomic engineering. However, most of us are familiar with the basic phenomena that are exploited in the steam engine and the ordinary explosives. Atomic explosives already influence international relations more profoundly than ordinary explosives, and it is not impossible that atomic energy may compete with our present sources of power in a few years. Tomorrow the basic facts about atomic power will be common knowledge. Even today a closer acquaintance with these facts may increase our foresight and help us to form our opinions both on questions of internal affairs and on those of foreign policies.

Atomic Reactions versus Ordinary Chemical Processes. The first question that one might ask concerns the special characteristics of atomic power. The combustion of 1 pound of coal suffices to raise the temperature of 700 pounds of water by 18 degrees Fahrenheit. But "combustion" of 1 pound of uranium would produce an equal temperature rise in 2 billion pounds of water. The same amount of energy is liberated when a pound of uranium explodes, whereas the explosion of 1 pound of nitroglycerin liberates only enough energy, when converted into heat, to raise the temperature of 150 pounds of water. What is the difference between the atomic process and our ordinary chemical reactions that makes the former so much more powerful?

The answer is that our ordinary chemical reactions affect the *arrangement* of the atoms, the "smallest building blocks of matter," but not their identity; atomic reactions change the *identity* of atoms. The burning of coal results in a breakup of the arrangement of the carbon atoms in the coal and the oxygen atoms in the air, out of which a new association of the carbon and oxygen atoms is formed. The chemist denotes the carbon atoms by C, the oxygen atoms by O. He would describe the burning of coal symbolically:

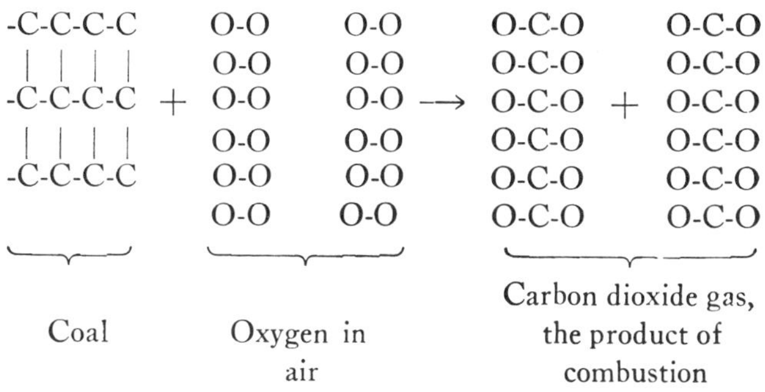

```
-C-C-C-C      O-O    O-O        O-C-O     O-C-O
 | | | |      O-O    O-O        O-C-O     O-C-O
-C-C-C-C  +   O-O    O-O  -->   O-C-O  +  O-C-O
 | | | |      O-O    O-O        O-C-O     O-C-O
-C-C-C-C      O-O    O-O        O-C-O     O-C-O
              O-O    O-O        O-C-O     O-C-O

  Coal        Oxygen in     Carbon dioxide gas,
                 air          the product of
                                combustion
```

Inasmuch as a chemical change such as the one above changes only the arrangement of the atoms, the number of atoms of a definite kind is the same before and after the reaction. There were twelve C atoms and twenty-four O atoms before the reaction, and there are twelve C atoms and twenty-four O atoms after the reaction. All that happens is that the C atoms are torn out of their lattice, the O atoms separated from their partners, and new unions formed between C and O atoms.

The difference between a fuel such as coal and an explosive such as nitroglycerin is that the nitroglycerin contains everything necessary for the reaction within itself, while the fuel needs another substance, namely air, for burning.

Atomic reactions are quite another matter. They change the atoms themselves. Thus the reactions which go on in an exploding atomic bomb are expressed in this way:

$$\text{U-235} \rightarrow \text{I} + \text{Y}$$

This is to say that uranium transforms into iodine

and yttrium, a rather rare element. (It can also break up into many other pairs of elements.) This change of one atomic species into others is contrary to the principles of ordinary chemistry. It is an effect that the medieval alchemists sought in vain for several centuries and that was realized only after their hopes were abandoned and the futility of their efforts raised to a general principle. This principle, now superseded, is called the principle of the immutability of elements. It holds in chemical but not in atomic processes.

None of this, of course, explains why the energy changes in atomic reactions are so much greater than energy changes in ordinary chemical reactions. Quite to the contrary, even front-line scientists puzzle as to the exact source of the atomic energy.

Einstein's famous equation, $E = mc^2$* tells us that in order to obtain the energy released in the reaction of the atomic bomb, we must subtract from the mass of U-235 the masses of I and Y and then multiply the remainder by the square of the velocity of light. This is a most useful rule that follows from a very fundamental relation. It does not tell us, however, why the mass of U-235 is greater (0.1 per cent, which is a great deal so far as mass differences go) than the combined masses of I and Y. From a general standpoint, it appears reasonable that a change that results in so fundamental an alteration of properties as the transformation of one element into another (or two others) should be connected with larger energy changes than the mere rearrangement of elements—and we must be content with this explanation.

Einstein's equation does tell us, however, how to calculate the energy released in any process if the masses of the atoms participating in the reaction are known. It tells us, for instance, that the energy released in the transformation of hydrogen into helium is about seven times greater (per unit weight of material reacting) than the energy released in the reaction of the atomic bomb (so-called "fission reaction"). It also tells us that the most powerful reaction of all is the one which has *no* final products. This is the so-called "annihilation reaction":

$$U \rightarrow$$

More will be said about these reactions later on. But it can be noted here that their initiation on a measurable scale still belongs in the dream world of scientists.

Still other atomic reactions occur spontaneously in the phenomenon known as radioactivity. This phenomenon occurs in many of the heavy elements found in nature, such as radium and thorium, and also in some artificially created forms of elements ordinarily stable. The iodine and yttrium, for example, that are products of uranium fission, are radioactive forms of ordinary stable iodine and yttrium. Radioactive atoms eject part of their substance and are thus changed into other elements. Sometimes the ejected particles are accompanied by a radiation known as gamma radiation, whose rays are similar to but more energetic and penetrating than X-rays.

The emission of the particles and rays proceeds at a rate determined by the probability of the occurrence of certain configurations within the nucleus of the atom. Nor can this rate be altered by such outside influences as heat or pressure. It is usually described by a quantity known as the half life, which is the time taken for half of any given quantity of material to disintegrate. At the end of one half life, only half of the original substance remains; at the end of two half lives only a quarter of it is left, and so on.

Isotopes and Isotope Separation. There is another difference between ordinary chemical reactions and atomic reactions that deserves consideration. It is a difference connected with the phenomenon of isotopes. Isotopes are forms of the same element: They behave so much alike in ordinary chemical reactions that it was, for a long time, an open question whether a mixture of two isotopes could be separated into its constituents.

Because isotopes are forms of the same element, they have the same chemical symbol. If one wants to distinguish them, one adds a number, denoting the approximate mass of the isotope, to the symbol of the element. U-235 is an isotope of the element uranium; U-238 is another, heavier isotope of the same element. Because isotopes behave so much alike in ordinary chemical reactions, it is not necessary to specify them in chemical processes. The burning of one isotope of carbon is so similar to the burning of the other isotope that one can speak simply about the burning of carbon.

* Energy equals mass times the velocity of light squared.

Not so in atomic reactions. Isotopes differ as much in their behavior in atomic processes as do wholly different elements in ordinary chemical reactions. Thus, for instance, it is so much more difficult to induce the reaction of the atomic bomb in U-238 than it is to induce it in U-235 that U-238 cannot be used in atomic bombs.

So much should emphasize at once the importance and the difficulty of the process of isotope separation. If one wants a very reactive substance, it is usually necessary to select a particular isotope of an element. U-235 is such a substance. And yet to free this isotope from the other isotopes of the same element is a very difficult job because they all behave under ordinary conditions in a very similar way. The difficulty is the same which one would encounter if coal occurred in nature only mixed with some other substance, such as clay, and if clay looked and behaved in all physical processes so much like coal that it would be impossible to wash it away from the coal without washing away the coal at the same time, or to separate the two by any other process.

Why Were Atomic Reactions Not Discovered Before? It may be asked fairly at this point why atomic reactions remained undetected for such a long time if they liberate such huge amounts of energy? Why are they not more evident in everyday life?

If we want to burn coal, we have to warm it up first to several hundred degrees temperature. Below this "temperature of inflammation", the combustion, if it proceeds at all, proceeds only imperceptibly. It is natural that atomic processes that furnish so much more heat and energy should require for their initiation a great deal more preheating than coal does. Such high temperatures as are necessary for this process can hardly be realized on our planet with our very limited resources. Temperatures high enough for atomic reactions prevail, however, at the center of the stars and our sun, and the source of the sun's radiation is atomic energy. Since all our terrestrial energy ultimately derives from the sun's radiation, one can say that atomic energy does form the basis of our life and our sources of energy.

There is a substance that requires very much less preheating for inflammation than does coal—phosphorus. A match will burst into flame on very little rubbing. Fire remained undiscovered for such a long time because there is no free phosphorus in nature. Even if there had been any, it would have burned away accidentally long before man could have laid his hands on it.

There is an atom called "neutron" (the "zero" element), which can react with almost every other element at ordinary temperatures. However, there are under ordinary conditions no neutrons in nature. The neutron was discovered only a few years ago (1932) by the British physicist Chadwick. The reason for the scarcity of neutrons is the same as the reason for the scarcity of phosphorus: Any neutrons that might be formed accidentally would react so soon with other elements that there are always very few of them around—very few indeed.

This, then, is the reason that we knew so little about atomic reactions until recently, that we succeeded only a few years ago in initiating them on a large scale: Reactions that do not involve neutrons need extremely high temperatures for their initiation; neutrons, on the other hand, are so reactive that they attach themselves to other atoms and thus cease to exist.

The Chain Reaction. Before 1939, most physicists believed on the basis of the facts just related that the use of atomic energy (strictly speaking it is nuclear energy because the changes that take place in atomic processes affect the nucleus of the atom) on an appreciable scale was far in the distant future. The neutrons that they succeeded in producing with great difficulty were all absorbed almost as soon as they were liberated, and atomic reactions with other elements could be induced artificially only by making use of a few very fast "hot" particles in an otherwise cold system. These fast particles were either the products of "radioactive substances" or were artificially generated in complex instruments such as the cyclotron or the Van de Graaff generator.

In 1939, two German scientists, Hahn and Strassman, discovered an atomic reaction, which was induced, as many other reactions are, by neutrons at ordinary temperature. The neutron that induced the reaction was absorbed in the process, as it is in all other processes which it induces. The decisive difference here, however, was that this atomic reaction also *produced* neutrons. It is clear that if the number of neutrons produced in the process is larger than the number of neutrons absorbed in it, it becomes possible not only to keep the reaction going at ordinary temperatures but also to obtain an abundant source of neutrons. What Hahn and Strassman had discovered was the fission process. It was mentioned back at the beginning of this chapter although its equation was not given fully. It is

$$\text{U-235} + \text{neutron} \longrightarrow \text{I} + \text{Y} + \text{N neutrons}$$

N denotes the number of neutrons yielded by one fission. The I and Y are called fission "fragments" because they are the fragments into which the U-235 breaks up. I and Y are not the only two elements into which the U-235 can "fission"; there are many other pairs of elements into which it can break up.

The important point in the above reaction is than N is larger than one. In fact, it is about two. This fact can be made use of in two ways if you have a lump of U-235 or of any other fissionable substance—that is, a substance that breaks up when absorbing a neutron.

The Bomb. Given a lump of U-235 or another fissionable material, you can add a neutron to it. This will then react with the U-235, giving two neutrons. If both of these neutrons are left to react with the U-235, they will produce four neutrons in the second generation. If all these react with the U-235 there will be 8 in the third generation, 16 in the fourth, about 1,000 in the tenth, 1 million in the twentieth, 1 billion in the thirtieth generation, etc. The processes induced by the neutrons of one generation will furnish the neutrons of the next generation, each generation being about twice more populous than the preceding one. This succession of events will continue either until all the U-235 is consumed and replaced by fission fragments and neutrons, or until the bomb has flown apart. For the system just described is a bomb, the atomic bomb.

The fission fragments of the reaction in the bomb have velocities which correspond to about a trillion degrees temperature, and the energy generated when a pound of U-235 has undergone fission is sufficient to raise the temperature of an air sphere of more than a half mile diameter to the boiling point of water. Actually, the destruction caused by such an explosion may extend over a greater area than that.

The life cycle of the neutron generations in a bomb is not much longer than a billionth of a second and the whole process described above can be over in a millionth of a second. The prime difficulty in the construction of the bomb is to keep together the lump of U-235 in spite of the huge energy evolution and to see to it that all or nearly all of the neutrons are absorbed by the U-235.

The Neutron Generator. The second way to make use of a mass of fissionable material is to let the number of neutrons increase only to a rather high but predetermined level and to stop their increase once that level is reached. The increase can be halted, for instance, by introducing into the system some foreign material that absorbs about one-half of all the neutrons being produced. If this is done, only one-half of all the neutrons of a generation will induce fission in the U-235. Since the number of neutrons in any generation will be twice greater than the number of fissions, the number of neutrons in each generation will be the same. In other words, the reaction will proceed at a steady rate that may be high or low, depending on the level at which the further increase of the neutrons is stopped. In all practical cases, the level is so low that it takes many weeks before an appreciable fraction of the U-235 is used up. This contrasts with the ten-millionth of a second in the bomb.

Running the chain reaction in this way will have two effects: (1) The fission processes that go on at a steady rate will produce a certain amount of heat that can be directed to useful purposes; (2) concurrently and inseparably, neutrons will be available to be absorbed in whatever one chooses to use to stop their excessive multiplication.

The second point is as important as the first one. Most nuclei become radioactive when absorbing a neutron, and it is therefore possible to make about one radioactive atom for every atom of U-235 that is used up. In this way a wide variety of radioactive atoms can be manufactured, because the chain reaction can be controlled by absorbing the excess neutrons with almost any of the ninety-two known elements. The last point illustrates the immense value of neutrons: One can induce an atomic process with every neutron and make, for instance, a radioactive atom out of almost any atom and a neutron. Another reason to avoid wasting the U-235 in a bomb! All the neutrons that the U-235 is able to produce are uselessly lost after the explosion.

The Plutonium Factory. The two preceding sections have a very great "if" to them. In order to make a bomb or build the neutron generator, you have to have first a rather large amount of fissionable material. Of course, it is possible to make pure U-235 by separating the two isotopes of uranium. However, if this were the only method of producing fissionable material for the neutron generator, the neutrons would remain very expensive indeed. The whole process can be made much cheaper if you can use natural uranium, that is, the mixture of U-235 and U-238.

This is indeed possible if the U-238 is used as the substance to prevent the increase of the number of neutrons. But doing this disposes of any choice as to the element that shall be permitted to absorb the neutrons: It is the U-238. What one has is not a neutron generator but a neutron generator and neutron consumer combined. The U-235 is the source of the neutrons, the U-238 their destination. It would seem that there is little gain except for the energy generated by the reaction.

However—and this is the "joker"—the product of the reaction of U-238 and the neutron is a new substance, U-239, which goes over, by a spontaneous radioactive "decay," into a new element called plutonium. And plutonium is fissionable too. As a result, it can be used either in a bomb or in another neutron generator. The choice of the neutron absorber, forced on us by the fact that the U-235 is always mixed with U-238, is not such a bad one. In fact it could hardly be better.

The plutonium factory just described is an unusual factory, indeed. It makes plutonium but while it makes it, it also generates energy. This is the energy of the process: U-235 + neutron = I + Y + N neutrons and of similar fission processes which furnish the neutrons.

The plutonium manufactured in the plutonium factories in the state of Washington is, by the way, the first new element ever manufactured by man in sizable quantities. And the plutonium can be obtained in the factory very much cheaper than pure U-235 can be obtained by isotope separation, to say nothing of the energy generated in the fission process. An almost unlimited amount of plutonium could be made within not too many years, enough for a very large number of bombs—or for peaceful, socially useful applications. Here we have a choice.

Because natural uranium can be made chain-reacting, it might be thought that it could be made to explode. It cannot: the neutron multiplication is not sufficiently rapid in natural uranium. The U-238 automatically controls the chain reaction, that is, absorbs so many neutrons that the population of the successive generations hardly increases. In fact quite a few tricks are called for to provide even a small increase in the population of successive generations or even to avoid a decrease. The most important of these tricks is to moderate the neutrons, to decrease their velocities from the high value that they have when they are expelled during fission to a fraction of that value (from 10,000 miles per second to about 1 mile per second). But in spite of all tricks the neutron multiplication in a system using natural uranium cannot be made fast enough for a bomb.

Other Atomic Reactions. Our excursion into the field of atomic physics brought us to a number of passes from which we could see other roads leading toward the realization of other atomic reactions. Two of these were specifically mentioned: The reaction between hydrogen isotopes (H^2), and the annihilation reaction. Both of these are virtually capable of yielding more energy per unit weight than does the fission reaction—the latter reaction about a thousand times more. What about them?

Not much. While, as we saw, the fission reaction is ready to be exploited on a large scale for better or worse, there is at present no reason to believe that any other atomic reaction could be exploited in the near future. It is suggestive to attempt to use the fission reaction to produce high temperatures and kindle by it other reactions, just as the fire of phosphorus is used to kindle fire in other substances. It has even been suggested that the atmosphere or the seas might be "set on fire" by fission bombs. At present there is no reason to fear this; the ignition of the atmosphere or the seas is pure speculation and I believe bad speculation. As for the annihilation reaction, it has hardly been observed, if at all, in the laboratory. Of course, we must guard against overconservatism, as the people who scoffed at the idea of a uranium chain reaction can testify. It may be a sound judgment, however, to believe that other, perhaps biological, discoveries of equal potency for bad or good may be made before we have to face atomic reactions of a fundamentally different nature than those that we can exploit now.

Should we be sorry that the side roads do not seem to lead anywhere at the present? I do not believe so. The amount of energy that available sources are able to furnish is so ample that we are not in need of other or more abundant sources. Present sources are able to fulfill all reasonable—and some unreasonable—needs for energy.

Chapter 4

GALE YOUNG, *formerly head of the mathematics and physics department at Olivet College, joined the Metallurgical Laboratory in Chicago in March, 1942, to work with the theoretical group there on problems connected with the design of the Hanford plutonium plant.*

The New Power

by GALE YOUNG

THE world may not have been familiar with atomic bombs prior to Hiroshima, but it has for years been talking about something called "atomic power." It has often been said that man would someday "unlock the energy of the atom," and the attainment of this goal has been frequently described in fiction. *The World Set Free,* written in 1914 by H. G. Wells, is such a book. There may be some doubt, however, about the current appropriateness of its title.

So it is that atomic power finally makes its appearance before a world that has been more or less expecting it for a considerable length of time. Indeed, as a look through the files of *Astounding Science Fiction* or some similar magazine will show, the real scientists still have a long way to go before they can hope to equal exploits that are commonplace to their fictional brother scientists. In comparison with these more fanciful accounts the real development may seem a bit disappointing. In the destructive-bomb department the new energy appears to be a prompt and thorough success, but as the much heralded servant of mankind, atomic power still leaves considerable to be desired.

It is the purpose of this chapter to provide some account of atomic energy as a source of useful power, although it is not possible to give anything like a full discussion at this time. There has been little time during the war to consider peaceful applications in this field, and much work will be needed before a proper perspective is attained.

Fission as a Power Source. Several kinds of heavy nuclei, for instance, uranium and plutonium, can be made to undergo the process of fission. The fission of 1 pound of any of these substances gives energy equal to that obtained by burning about 1,400 tons of coal or 900 tons of gasoline, or by exploding 13,000 tons of TNT. According to the Einstein mass-energy rule ($E = mc^2$), a pound of matter is equivalent to the energy obtained in burning 1.5 million tons of coal. One sees, therefore, that the fission process changes only 1/1,100 of the available mass into energy; evidently the production of atomic power is still a highly imperfect business.

As explained earlier, in undergoing fission the nucleus of the atom splits into two pieces of comparable size that fly apart at great speed. Most of the energy released goes into the kinetic energy of these moving fragments, and as these are slowed down by the material through which they move this energy is turned into heat. These fragments carry an electrical charge, and if they could be stopped by an electrical field the energy could be obtained in electrical form instead of heat; this can be imagined in principle, but it appears that in practice heat will be obtained.

Besides the two large fragments just discussed, the fission explosion also emits gamma rays and fast neutrons. Because of their lack of electrical charge the neutrons can pass through large thicknesses of solid matter; they thus constitute another kind of penetrating "radiation" in addition to the gamma rays. Since both neutrons and gamma rays are dangerous to living tissue, both must be shielded against in operating a fission power source. In this respect a fission source is even worse than a radium source; each gives out several per cent of its energy in the form of dangerous penetrating radiation, but the mixed nature of the fission radiation makes it harder to cope with and requires greater shield thicknesses. For example, lead is effective against gamma rays, but neutrons romp through it with ease; water is good against neutrons since it slows them down and renders them easily absorbable, but it has little stopping power for gamma rays. A thin lightweight shield would be one of the most valuable inventions in the atomic-power field; unfortunately this just does not seem to be in the cards.

After the two fission fragments slow down, they gather electrons around them and become the nuclei of two new atoms. These resulting nuclei are unstable and undergo successive radioactive transformations in the course of which they emit particles and gamma rays much as does radium; this radioactivity dies away very slowly. Thus, while a fission reactor may be free of radioactivity when first assembled, as soon as it starts to operate the strongly active fission fragments begin to accumulate inside it and make shielding and cooling necessary even when the reaction is not going on.

Atomic Power Plants. The fission chain reaction shows a striking critical-size effect. A small isolated lump of material will not react, no matter what is done to it. As the size of the lump is increased it will, assuming it is a proper kind of material, eventually begin to react. If more material is added the reaction will continue to increase; if material is removed the reaction will decrease and die out. Evidently an operating unit must be just the right size; if too small it will not run at all, and if too large it tends to run away. The reaction can be controlled by moving a small piece of material into and out of the system.

The reason for this behavior lies in the chain nature of the reaction. The explosion of one nucleus releases neutrons that move away and cause another nucleus somewhere else to explode, and so on. Not every neutron ejected in a fission goes on to cause another fission. Some will be captured uselessly and some will escape from the system before they are captured at all. The smaller the structure, the greater the fraction of neutrons that escape. For the reaction to proceed at a constant rate, the structure must be just large enough so that, of the N neutrons emitted in a fission, $N - 1$ neutrons escape or are uselessly captured, while the remaining one neutron produces another fission.

Thus a reactor will not run unless it contains a certain amount of material. The amount of valuable fissionable material needed can be decreased by a number of tricks, but it can never be made very small. Therefore, any notion to the effect that an engine can be run on a small capsule of atomic fuel is completely without foundation.

One of the tricks for decreasing the amount of fissionable material needed is to mix it with light atoms, such as carbon or beryllium or hydrogen, which slow down or moderate the fast neutrons and make it easier for them to cause new fissions. In some cases the difference is not merely in the amount of material required, but is the difference between running and not running at all. Thus, it is impossible to make a chain reaction with natural uranium alone, no matter how much is available; but the reaction can be made to go with natural uranium plus any of several existing moderator materials. In the early days of the project there were much piling up and unpiling of materials in experimental work to determine the best ways of arranging uranium in the moderator; this has led to the name "piles" for chain-reacting units, even when they are precision-engineered structures.

Atomic fuel, for instance uranium, generates heat inside itself in a properly constructed pile without any need for air or any other chemical participant. If the atomic reaction proceeds at a rapid rate, this heat must be removed or the pile will melt. At the Hanford plutonium plants a product of the reaction other than the heat was wanted, so the heat was carried away by cooling streams which emptied into the Columbia River. If the heat is wanted for the production of power, it must first be removed from the pile and then introduced into a heat engine. Apparently atomic power will have to function through heat engines of more or less standard design; there is no reason, therefore, to expect any marvels in the way of performance. While a few pounds of uranium may have enough releasable energy to drive the *Queen Mary* across the ocean, it is not likely to do this via a few pounds of power plant.

Most of the existing chain reactors or piles use natural uranium with a graphite moderator. They consist of large graphite blocks pierced by regularly spaced parallel holes in which somewhat smaller uranium rods are centered. Cooling streams flow along annular channels about the rods, picking up the heat from the rods and carrying it away. The low-power pile at Oak Ridge, Tennessee, uses air cooling, and the higher power piles at Hanford, Washington, use water cooling; in both cases the cooling fluid is drawn from the environment, passed through the pile, and discarded back to the environment. In both cases the uranium rods are jacketed in aluminum to keep the coolant from touching them, and in the Hanford plants another layer of aluminum is used to keep the water from the graphite.

Some arrangements that might give mechanical power output are indicated in the following sketches. Figure 1 shows an obvious modification of the Oak Ridge type of pile into a gas-turbine plant. Since air becomes radioactive when exposed to neutrons, it has to be disposed of suitably, say by discharge from a tall stack; and the ability of the stack to get rid of active air in a safe manner may limit the power output of the plant. Charged lines in the drawing indicate where the air is active; it is seen that the turbine is exposed to, and must be operable under, radioactive conditions. For the plant to operate it is necessary that the pile should heat the air to quite high temperatures; the existing piles cannot do this because both graphite and aluminum are attacked by air at elevated temperatures. The development of piles for high-temperature operation is one of the main problems to be solved before useful power can be obtained.

Passage to a closed-cycle system, as in Fig. 2, would eliminate the stack-discharge problem and permit the use of a nonoxidizing working substance such as helium; the use of helium also helps to reduce the radioactivity in the pile circuit. However, a heat exchanger is now needed, and the compressor is now in the radioactive region. The diagram shown can, of course, be elaborated to include regenerators, compressor intercooling, etc., to give higher efficiency.

Another possible arrangement is shown in Fig. 3, in which the pile coolant gives its heat to a more or less standard power plant boiler. Since a liquid-vapor power plant can run at considerably lower

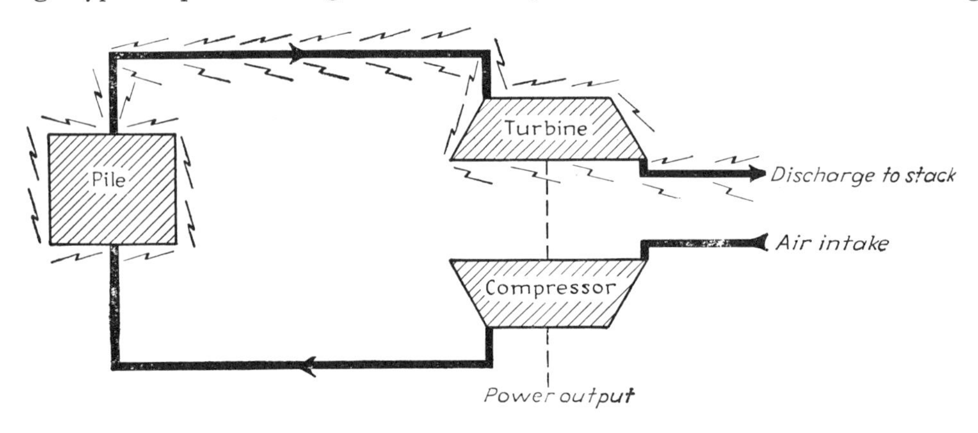

Fig. 1

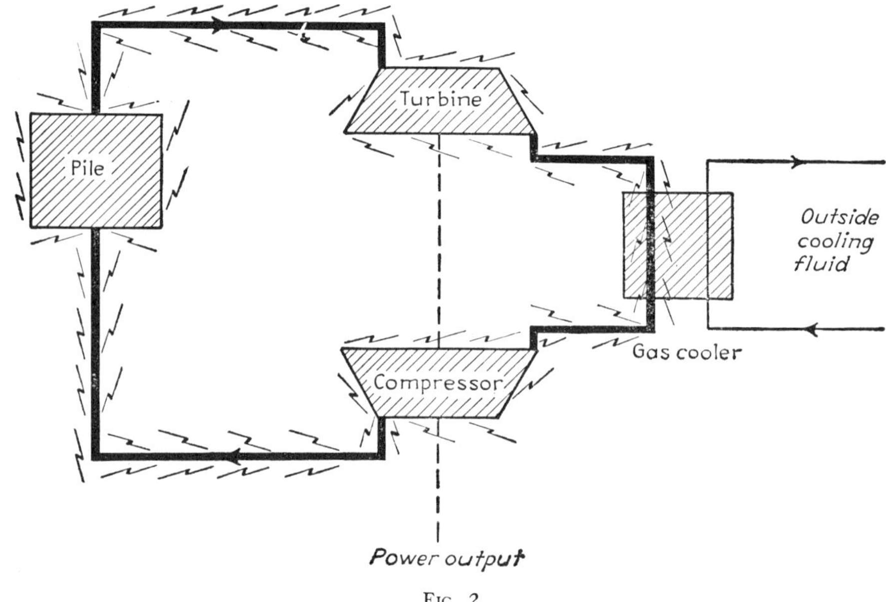

Fig. 2

temperatures than a gas turbine plant, it may be possible to get power in this arrangement without such high temperatures in the pile.

The present Hanford plants with water cooling operate at very low temperature. Development of a liquid-metal coolant, such as bismuth or sodium, would permit higher temperatures, and power might be obtained as in Fig. 3.

It would seem, therefore, that we are not in dire need of more energy sources.

Actually, of course, we are interested not just in energy generally, but in energy that will work what, when, where, and how we want it to work. Sunlight falling all over the ground at better than a horsepower per square yard will not help a man to move a rock unless he gets together a lot of equipment by means of which he can put the solar energy to work. It would, in fact, take an unusually expensive installation of mirrors and boilers to move a rock by means of sunshine.

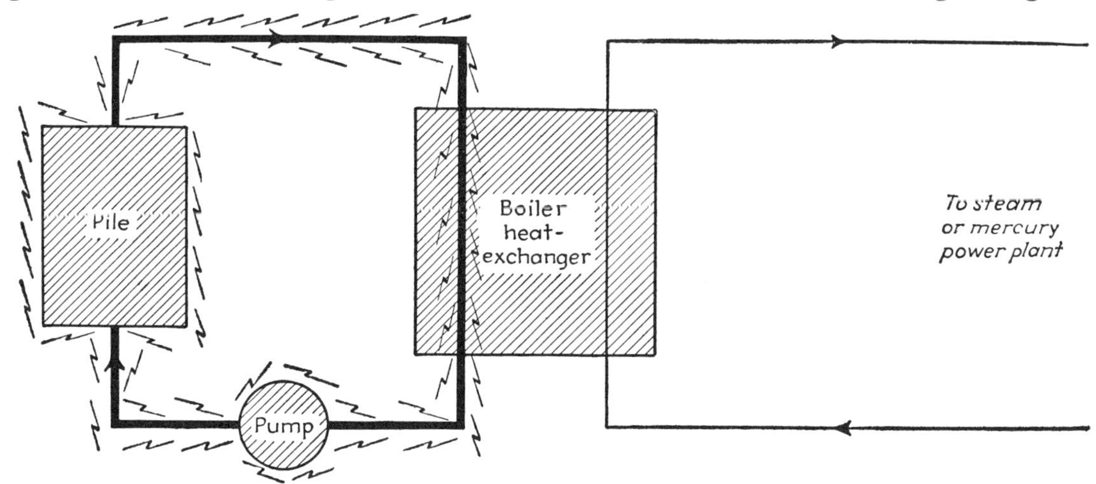

FIG. 3

Arrangements in which the pile directly vaporizes light or heavy water to make steam for a power plant are also conceivable. Among the difficulties in this case are the need for the pile structure to withstand boiler pressures, and the adverse effect of bubble formation upon the stability of pile operation.

Economics and Prospects of Atomic Power. There has been considerable speculation about the use of atomic energy for peaceful purposes. All manner of predictions and estimates have been heard, ranging from ridiculous optimism to perhaps undue pessimism.

Why, one may ask, is there so much excitement about a new source of power? We live in a world in which power is going to waste all the time in the form of sunlight, winds, water falls, tides, and so on. At our present rate of fuel consumption, the coal supply will last a few thousand years. This is a period considerably greater than technological civilization has existed in the past and, if some current events are an indication, considerably greater than it can expect to exist in the future. Petroleum resources are running lower, but it appears that gasoline can be made from coal in satisfactory fashion, or that alcohol produced from crops each year can be used as liquid fuel.

Much the same is true of the other "free" energy sources; they are generally not worth the investment required to put them to work. Only water power among them finds any important use in direct power production.

Fuels such as coal and gasoline are not free; labor and equipment are needed for gathering, processing, and transporting them, and these items are lumped into the basic cost of the fuel. After the fuel arrives, more effort and equipment are involved before power is obtained in the desired form, all of which represents a utilization cost.

An everyday example of these two kinds of cost may be obtained by comparing the utilization cost for the family automobile—depreciation, interest, maintenance, insurance, repairs, license, garage rent, etc.—with the amount spent for gasoline. In commercial truck operation about 0.1 of the total expense is for gasoline. Another example is given in Table 1. It is seen that fuel is by no means the major part of the total cost; free coal and gasoline would be very nice, but probably would not usher in any "new era," and we would still have most of our power bill to pay.

TABLE I

COST OF ELECTRICAL ENERGY TO AVERAGE URBAN CONSUMER*

Type of Cost	Cents per kw-hr
Generating cost, including fuel	0.47
Fixed charges (interest, depreciation, etc.) on generating station	0.78
Transmission and substation operation	0.16
Fixed charges on transmission and substation equipment	0.33
Operation cost of distribution lines from substation to consumer	0.29
Fixed charges on distribution equipment	1.45
Administration, bookkeeping, reading meters, service calls, etc.	1.80
Total cost at customer's meter	5.28

* From Barnard, Ellenwood, and Hirshfeld, *Heat-power Engineering*, p. 1054, 1933.

Cost figures on atomic fuel have not been released as yet, and no power plants have been developed. It is not possible, therefore, to make any meaningful estimates of how the total cost (fuel + utilization) would compare with the cost of other power sources.

It may be interesting to note prewar quotations of unpurified natural uranium at about $1.80 a pound; this is around $250 for a pound of unseparated U-235 equivalent energywise to about $8,000 worth of coal or $30,000 worth of gasoline. If the necessary processing, such as purification or reduction to metal or isotope separation, does not prove too expensive, uranium may be able to compete so far as fuel cost is concerned.

With respect to utilization cost, we may recall that an atomic power plant involves machinery comparable to that of ordinary steam or gas turbine stations except that it has a pile in place of the burner or combustion chamber and is complicated by radiation and shielding problems. It is doubtful, therefore, that the cost and maintenance of an atomic plant can be lower than that of an ordinary fuel plant, and it may in fact be considerably higher.

The United States average rate of mechanical-power production is roughly given in Table 2. Much of the stationary power plant output is in electrical energy; the electrical production is about two-thirds from fuel burning stations, and one-third from hydroelectric plants.

Atomic power is unable to compete in the important automobile field because of the large weight (dozens of tons) of shielding required around a pile and also because of the large cost represented by the amount of fissionable material needed to make a chain-reacting unit. Large locomotives could conceivably carry the necessary weight of shielding and might be regarded as a borderline case. Ships and stationary power stations are definite possibilities for atomic power; the economics of fuel and plant costs will, of course, need to be taken into account, as well as the relative health and operating hazards involved.

Average heat outputs from mineral fuel are indicated in Table 3. If the world total were

TABLE 2

AVERAGE MECHANICAL OUTPUT RATES

Power Plant	Million kw.
Autos, airplanes, etc.	25
Locomotives	7
Ships in trade	3
Stationary power plants	25
Total	60

TABLE 3

AVERAGE HEAT OUTPUT RATES

	U. S., million kw.	World, million kw.	U. S. per cent of world
Coal	500	2,000	25
Petroleum	300	500	60
Natural gas	100	110	90
Total	900	2,610	35

obtained from coal alone, about 3 billion tons per year would be required; this would use up the estimated coal reserves in about 2,700 years. About 0.4 of United States fuel use is for production of mechanical power, 0.2 for nonindustrial heating, and 0.4 for industrial heating (only a fraction of the energy used for production of mechanical power is actually converted into mechanical output). Thus, a considerable fraction of our fuel is used for heating rather than for power purposes. Buildings could be heated by atomic energy without high pile temperatures. And atomic energy plants, to mention one pleasing virtue, would not discharge any smoke into the atmosphere.

A specific advantage of atomic power is the small weight of the fuel itself. Thus, once the plant is installed, heating or power service may be feasible in remote locations where transportation costs would make the use of ordinary heavy fuels impractical. It is perhaps conceivable that a country without coal or oil resources might be able to maintain a power economy based on electricity generated in atomic power stations. However, considerable study by power engineers, economists, and others will be needed before we get beyond the level of mere speculation on such topics. Without answers to numerous such questions, the problem of atomic-energy control is apt to continue confused; and it is to be hoped that information will soon be released so that such studies can proceed.

Fuel weight is also of importance in mobile units. If the weight of coal or oil carried by a ship exceeds the weight of the shielding required to convert to atomic power, then total weight could be saved by so converting. Atomic-powered ships or, if such are ever possible, atomic-powered airplanes should be able to cover great distances without refueling.

Another advantage of atomic power is that it does not require oxygen and does not give off combustion gases; in some cases, this may justify considerable trouble in trying to cope with the disadvantages. We have thus the possibility of power plants functioning in confined places, underwater in submarines, underground, or, if various difficulties can ever be overcome, outside the earth's atmosphere in space ships.

There is a great deal of uranium in the earth's crust, but it is not known how much of this will be accessible for use. The energy present in rich uranium deposits known before the war is, as indicated in Table 4, negligible in comparison with the

TABLE 4

MISCELLANEOUS ESTIMATES

Population of earth	2 billion
Population of U. S.	0.13 billion
Energy rate per capita	7,000 watts for U. S. 1,300 watts for world
Metabolic rate of human body	100 watts resting 300 watts in brisk walk
Total energy used by man	3×10^9 kw.
Solar energy hitting earth	1.7×10^{14} kw.
Estimated world coal resources	8×10^{12} tons
Estimated U-235 in earth's crust	2×10^{12} tons
Sunlight equivalent of coal resources	15 days
Sunlight equivalent of total U-235	30,000 years
Sunlight equivalent of U-235 in known deposits	3 minutes

energy in known coal deposits; indeed, if these were our total uranium resources, it would be unwise to waste them in any use for which more abundant ordinary fuel could suffice. Doubtless there is now quietly underway a world-wide buried-treasure hunt such as has never been seen before.

Chapter 5

J. R. Oppenheimer before the war was professor of physics at the University of California and there led the most important school for theoretical physics in the United States. During the war he was in charge of the Los Alamos, New Mexico, laboratory. He is now at California Institute of Technology.

The New Weapon: *The Turn of the Screw*

by J. R. OPPENHEIMER

THE release of atomic energy constitutes a new force too revolutionary to consider in the framework of old ideas. . . . President Harry S. Truman, in his message to Congress on Atomic Energy, October 3, 1945.

In these brave words the President of the United States has given expression to a conviction deep and prevalent among those who have been thinking of what atomic weapons might mean to the world. It is the conviction that these weapons call for and by their existence will help to create radical and profound changes in the politics of the world. These words of the President have often been quoted and for the most part by men who believed in their validity. What is the technical basis for this belief? Why should a development that appeared in this past war to be merely an extension and consummation of the techniques of strategic bombing be so radical a thing in its implications?

Certainly atomic weapons appeared with dramatic elements of novelty; certainly they do embody, as new sources of energy, very real changes in the ability of man to tap and to control such sources, very real differences in the kind of physical situation we can realize on earth. These promethean qualities of drama and of novelty, that touch so deeply the sentiments with which man regards the natural world and his place in it, have no doubt added to the interest with which atomic weapons have been regarded. Such qualities may even play a most valuable part in preparing men to take with necessary seriousness the grave problems put to them by these technical advances. But the truly radical character of atomic weapons lies neither in the suddenness with which they emerged from the laboratories and the secret industries, nor in the fact that they exploit an energy qualitatively different in origin from all earlier sources. It lies in their vastly greater powers of destruction, in the vastly reduced effort needed for such destruction. And it lies no less in the consequent necessity for new and more effective methods by which mankind may control the use of its new powers.

Nothing can be effectively new in touching the course of men's lives that is not also old. Nothing can be effectively revolutionary that is not deeply rooted in human experience. If, as I believe, the release of atomic energy is in fact revolutionary, it is surely not because its promise of rapid technological change, its realization of fantastic powers of destruction, have no analogue in our late history. It is precisely because that history has so well prepared us to understand what these things may mean.

Perhaps it may add to clarity to speak briefly of these three elements of novelty: (1) atomic weapons as a new source of energy, (2) atomic weapons as a new expression of the role of fundamental science, and (3) atomic weapons as a new power of destruction.

As a New Source of Energy. The energy we derive from coal and wood and oil came originally from sunlight, which, through the mechanisms of photosynthesis, stored this energy in organic matter. When these fuels are burned, they return more or less to the simple stable products from which, by sunlight, the organic matter was built up. The energy derived from water power also comes from sunlight, which raises water by evaporation, so that we may exploit the energy of its fall. The energy necessary to life itself comes from the same organic matter, created by sunlight out of water and carbon dioxide. Of all the sources of energy used on earth, only tidal power would appear not to be a direct exploitation of the energy radiated by the sun.

Solar energy is nuclear energy. Deep in the interior of the sun, where matter is very hot and fairly dense, the nuclei of hydrogen are slowly reacting to form helium, reacting not directly with each other but by a complex series of collisions with carbon and nitrogen. These reactions, which proceed slowly even at the high temperatures that obtain in the center of the sun, are made possible at all only because those temperatures, of some twenty million degrees, are maintained. The reason they are maintained is that the enormous gravitational forces of the sun's mass keep the material from expanding and cooling. No proposals have ever been made for realizing such conditions on earth or for deriving energy on earth in a controlled and large-scale way from the conversion of hydrogen to helium and heavier nuclei.

The nuclear energy released in atomic weapons and in controlled nuclear reactors has a very different source, which would appear to us now as rather accidental. The nuclei of the very heavy elements are less stable than those of elements like iron, of moderate atomic weight. For reasons we do not understand, there are such heavy unstable elements on earth. As was discovered just before the war, the heaviest of the elements do not need to be very highly excited to split into two lighter nuclei. It is surely an accident, by all we now know, that there are elements heavier than lead in the world, and for lead no practical method of inducing fission would seem to exist. But for uranium the simple capture of a neutron is sufficient to cause fission. And in certain materials—notably U-235 and plutonium—enough neutrons are produced by such fission, with a great enough chance that under proper circumstances they can cause other fissions in other nuclei, so that the fission reaction will build up in a diverging chain of successive reactions, and a good part of the energy latent in the material actually will be released.

To our knowledge such things do not happen except in the atomic weapons we have made and used and, in a somewhat more complex form, in the great reactors or piles. They do not, to our knowledge, happen in any other part of the universe. To make them happen involves specific human intervention in the physical world.

The interior of an exploding fission bomb is, so far as we know, a place without parallel elsewhere. It is hotter than the center of the sun; it is filled with matter that does not normally occur in nature and with radiations—neutrons, gamma rays, fission fragments, electrons—of an intensity without precedent in human experience. The pressures are a thousand billion times atmospheric pressure. In the crudest, simplest sense, it is quite true that in atomic weapons man has created novelty.

As a New Expression of the Role of Science. It would appear to be without parallel in human history that basic knowledge about the nature of the physical world should have been applied so rapidly to changing, in an important way, the physical conditions of man's life. In 1938, it was not known that fission could occur. Neither the existence nor the properties nor the methods of making plutonium had been thought of—to the best of my knowledge—by anyone. The subsequent rapid development was made possible only by the extremities of the war and the great courage of the governments of the United States and Britain, by an advanced technology and a united people. Nevertheless it made very special demands of the scientists, who have played a more intimate, deliberate, and conscious part in altering the conditions of human life than ever before in our history.

The obvious consequence of this intimate participation of scientists is a quite new sense of responsibility and concern for what they have done and for what may come of it. This book itself is an expression of that sense of concern. A more subtle aspect of it, not frequently recognized but perhaps in the long term more relevant and more constructive, is this: Scientists are, not by the nature of what they find but by the way in which they find it, humanists; science, by its methods, its values, and the nature of the objectivity it seeks, is universally human. It is therefore natural for scientists to look at the new world of atomic energy and atomic weapons in a very broad light. And in this light the community of experience, of effort, and of values that prevails among scientists of different nations is comparable in significance with the community of interest existing for the men and the women of one nation. It is natural that they should supplement the fraternity of the peoples of one country with the fraternity of men of learning everywhere, with the value that these men put upon knowledge, and with the attempt—which is their heritage—to transcend the accidents of personal or national history in discovering more of the nature of the physical world.

The injection of the spirit of the scientist into this problem of atomic weapons, in which it has been clear from the first that purely national ideas of welfare and security would doubtless prove inadequate, has been recognized, if not clearly understood, by statesmen as well as by scientists. The emphasis that has been given—in the statements of the President and in the agreed declaration of the heads of state of Britain, Canada, and the United States—to the importance of the re-establishment of the international fraternity and freedom of science is an evidence of this recognition. It should not be thought that this recognition implies either that collaboration in science will constitute a solution to the problems of the relations of nations, nor that scientists themselves can play any disproportionate part in achieving that solution. It is rather a recognition that in these problems a common approach, in which national interests can play only a limitedly constructive part, will be necessary if a solution is to be found at all. Such an approach has been characteristic of science in the past. In its application to the problems of international relations there is novelty.

As a New Power of Destruction. In this past war it cost the United States about $10 a pound to deliver explosive to an enemy target. Fifty-thousand tons of explosive would thus cost a billion dollars to deliver. Although no precise estimates of the costs of making an atomic bomb equivalent to 50,000 tons of ordinary explosive in energy release can now be given, it seems certain that such costs might be several hundred times less, possibly a thousand times less. Ton for equivalent ton, atomic explosives are vastly cheaper than ordinary explosives. Before conclusions can be drawn from this fact, a number of points must be looked at. But it will turn out that the immediate conclusion is right: Atomic explosives vastly increase the power of destruction per dollar spent, per man-hour invested; they profoundly upset the precarious balance between the effort necessary to destroy and the extent of the destruction.

The area destroyed by explosive is a better guide to its destructive power than the energy of the explosion. For an atomic bomb the area destroyed by blast increases with the two-thirds power, not proportionately to the energy release. So far as blast is concerned, an atomic bomb is perhaps five times less effective than the same equivalent tonnage delivered in block busters or smaller missiles would be. But in the strikes against Hiroshima and Nagasaki the effects, especially the anti-personnel effects, of heat were comparable to the blast effects. These effects increase proportionately in the area affected with the energy release of the weapon and thus become of greater importance as the power of the weapon is increased.

In this connection it is clearly relevant to ask what technical developments the future might have in store for the infant atomic-weapon industry. No suggestions are known to me that would greatly reduce the unit size of the weapons while keeping down the cost per unit of area devastated or reducing it still further. On the other hand proposals that appear sound have been investigated in a preliminary way, and it turns out that they would reduce the cost of destruction per square mile probably by a factor of ten or more, but they would involve a great increase in the unit power of the weapons. Such weapons would clearly be limited in application to the destruction of very major targets, such as greater New York.

A word may be in order concerning the specific effect of the nuclear radiations—neutrons and gamma rays—produced by the explosion. The novel character of these effects, and the fact that lethal effects kept appearing many weeks after the strikes, attracted much attention. But these radiations accounted, in Hiroshima and Nagasaki, for a quite small fraction of the casualties. It is probable that this would be true of future atomic weapons as well, but it is not certain.

No discussion of the economics of atomic destruction can be complete without some mention of possible countermeasures. One such countermeasure, which might have the effect of increasing the cost of destruction by eliminating targets large enough for the more powerful weapons, is the dispersal of cities and industry. More difficult to evaluate is the effect of greatly improved methods of interception on the carrier of the atomic weapon. This question is examined in some detail by Dr. Ridenour in Chapter 7. It may be said here that the present situation hardly warrants the belief that techniques of interception will greatly alter the cost of atomic destruction.

Although it would seem virtually certain that atomic weapons could be used effectively against combat personnel, against fortifications at least of certain types, and against naval craft, their disproportionate power of destruction is greatest in strategic bombardment: In destroying centers of population, and population itself, and in destroying industry. Since the United States and Britain in this past war were willing to engage in mass demolition and incendiary raids against civilian centers and did in fact use atomic weapons against primarily civilian targets, there would seem little valid hope that such use would not be made in any future major war.

The many factors discussed here, and others that cannot be discussed here, clearly make it inappropriate and impossible to give a precise figure for the probable cost and thus the probable effort involved in atomic destruction. Clearly too such costs would in the first instance depend on the technical and military policies of nations engaging in atomic armament. But none of these uncertainties can becloud the fact that it will cost enormously less to destroy a square mile with atomic weapons than with any weapons hitherto known to warfare. My own estimate is that the advent of such weapons will reduce the cost, certainly by more than a factor of ten, more probably by a factor of a hundred.* In this respect only biological warfare would seem to offer competition for the evil that a dollar can do.

It would thus seem that the power of destruction that has come into men's hands has in fact been qualitatively altered by atomic weapons. In particular it is clear that the reluctance of peoples and of many governments to divert a large part of their wealth and effort to preparations for war can no longer be counted on at all to insure the absence of such preparations. It would seem that the conscious acquisition of these new powers of destruction calls for the equally conscious determination that they must not be used and that all necessary steps be taken to insure that they will not be used. Such steps, once taken, would provide machinery adequate for the avoidance of international war.

The situation, in fact, bears some analogy to one that has recently, without technical foundation, been imagined. It has been suggested that some future atomic weapon might initiate nuclear reactions that would destroy the earth itself or render it unsuitable for the continuance of life. By all we now know, and it is not inconsiderable, such fears are groundless. An atomic weapon will not, by what we know, destroy physically the men or the nation using it. Yet it seems to me that an awareness of the consequences of atomic warfare to all peoples of the earth, to aggressor and defender alike, can hardly be a less cogent argument for preventing such warfare than the possibilities outlined above. For the dangers to mankind are in some ways quite as grave, and the inadequacy of any compensating national advantage is, to me at least, quite as evident.

The vastly increased powers of destruction that atomic weapons give us have brought with them a profound change in the balance between national and international interests. The common interest of all in the prevention of atomic warfare would seem immensely to overshadow any purely national interest, whether of welfare or of security. At the same time it would seem of most doubtful value in any long term to rely on purely national methods of defense for insuring security, as is discussed in greater detail in other parts of this book. The true security of this nation, as of any other will be found, if at all, only in the collective efforts of all.

It is even now clear that such efforts will not be successful if they are made only as a supplement, or secondary insurance, to a national defense. In fact it is clear that such collective efforts will require, and do today require, a very real renunciation of the steps by which in the past national security has been sought. It is clear that in a very real sense the past patterns of national security are inconsistent with the attainment of security on the only level where it can now, in the atomic age, be effective. It may be that in times to come it will be by this that atomic weapons are most remembered. It is in this that they will come to seem "too revolutionary to consider in the framework of old ideas."

* See General Arnold's estimate, page 27.

Chapter 6

GENERAL OF THE ARMY H. H. ARNOLD *was a member of the U. S. Army Air Forces from 1919 and Chief of the Air Staff from March, 1943, to February, 1946. This chapter is his last official public statement as head of the Air Forces.*

Air Force in the Atomic Age

by General of the Army H. H. ARNOLD

"WE are aware that the only complete protection for the civilized world from the destructive use of scientific knowledge lies in the prevention of war."

This sentence from the joint declaration issued by President Truman, Prime Minister Atlee, and Prime Minister King in November, 1945, is a blunt assertion by three great nations that destruction by air power has become too cheap and easy. This was true even before the creation of the atomic bomb; even then, mass air raids were obliterating the great centers of mankind. The consequence of this cheapness of destruction—especially as it has been multiplied manifold by the sudden, extensive, pulverizing force of the atomic bomb—is to make the existence of civilization subject to the good will and the good sense of the men who control the employment of air power. The greatest need facing the world today is for international control of the human forces that make for war.

Pending the establishment of such control, the mission of the Air Forces is the protection of the United States by the employment of air power. Under the provisions of the charter of the United Nations, moreover, the Air Forces must hold immediately available contingents for combined international enforcement action. And to carry out this responsibility, the Air Forces must maintain a program of preparedness giving the best possible protection. Let me examine the problem as it now confronts us.

The Economics of Air Power. The biggest change that atomic explosives have made in the nature of air power is to decrease the cost of destruction. Complete destruction of cities by incendiaries and high-explosive bombs had become an accomplished fact early in the war; the name of Coventry reminds us of that. In the course of the war, strategic bombing advanced enormously in effectiveness and efficiency and, in the attack on the Japanese Empire, had become highly profitable from a military point of view. But the atomic bomb by comparison dwarfs all other advances. A dollars-and-cents approach to this problem will clarify these facts.

The B-29 program represents the high point in the continually advancing technique of strategic destruction by air power. By the latter part of 1945, the Twentieth Air Force was destroying Japanese industrial cities at the rate of 1 square mile for 3 million dollars of the war budget. This cost includes all supporting ground organizations, both continental and overseas. It is our best estimate of the *running* cost. Large investments in expansion, left uncompleted because of Japan's surrender, have been excluded.

No official figures have been released on the cost of the atomic bomb; but for purposes of comparison let us use, as an unofficial estimate of the mass-production cost per bomb, the figure of about 1 million dollars cited by Dr. Oppenheimer in the preceding chapter. The cost of delivering the bomb, represented by the flights of the bombing plane and several weather and reconnaissance planes, would be about $240,000, bringing the total cost per bomb used to about $1,240,000. The Hiroshima bomb destroyed 4.1 square miles and the Nagasaki bomb 1.4 square miles, giving an average of 2.8 square miles per bomb. Hence, the cost of destroying a square mile by this means is less than half a million dollars. So atomic bombing is, at the least, six times more economical than conventional bombing.

The estimated sixfold increase in economy is based on most conservative estimates. In the case of Nagasaki the shape of the target area was such that much of the destructive power of the bomb was directed at empty fields. Furthermore, the Nagasaki bomb was only the third to explode in all history, and improvements will make later models more effective. To destroy a square mile of a modern city in the future will cost much less than half a million dollars.

The destruction of cities is not the aim of a strategic air force. The aim is to weaken the enemy's military strength and will to resist to the point where he can successfully be invaded by ground forces, as was Germany, or capitulates in the face of certain destruction, as did Japan. In Germany the decisive air blows were aimed at oil and transportation; and although the total value of the destruction in dollars might not have been so large as that for cities, the effect was to paralyze the whole German war machine and make land conquest practical.

With the decreased cost of destruction afforded by the atomic bomb, the effort required to knock out completely all phases of an enemy's war industry is within practical reach. Once more an argument based on dollars will illustrate the point. The Tokyo earthquake of 1923 destroyed 11,000 acres at an estimated loss of 2,750 million dollars; thus, each square mile destroyed represented a loss of 160 million dollars to the Japanese.* The value of a square mile of Tokyo was, if anything, less at the time of the earthquake than at the time of the destruction by the B-29's. Since the cost to us of accomplishing this destruction was running at the rate of about 3 million dollars per square mile, the B-29 attacks were profitable by a factor of about fifty; that is, the cost to Japan was fifty times the cost to us. Although the figures just quoted give a reasonably true picture of the relative war costs to the two nations, it is necessary to point out that not all of the destruction to Japan had significance in impeding her war effort. A large fraction of the 160 million dollars per square mile of destruction represented personal property and institutions not needed in war production. However, the remaining small fraction of the damage (about one-eighth), which corresponded to strategic loss of war effort, was about six times greater than the cost to us of producing the damage. So far as the ultimate cost of the war to civilization is concerned, the value of the total destruction is of the most significance, and the economy of air power in producing such destruction is fifty-fold, as outlined above. But with the advent of atomic explosives, destruction will be at least six times more economical.† Thus, in a future war every dollar spent in an air offensive is expected to do more than $300 worth of damage to the enemy.

The numbers 50 and 300 are a tremendous argument for a world organization that will eliminate conflict by air power. These numbers, although they are only rough estimates, convey in compact form the grim fact that destruction is now too cheap, too easy. They represent a new phase in warfare. In the past, when one nation attacked another on the ground, the loss to each was limited by the forces and resources engaged. Now the limit is set only by the total resources exposed to atomic bombs. In the past a war might consume the income of a nation for some years; in the future it will also consume its capital. Dollars have been used as a measure of value because they are familiar and specific. It is hardly necessary to stress the arguments that can be developed in terms of lives, freedom from want, and the other values of civilization. In every case the conclusion will be the same: destruction by air power will be too easy. No effort spent on international cooperation will be too great if it assures prevention of this destruction.

The Growth of Air Power. To gain perspective on this subject, let us first consider the growing effectiveness of air power up to the present. In the First World War, air power played a negligible strategic role; its purpose was primarily tactical. In the European phase of the Second World War, the strategic effect was decisive and crippled the war machine to the point that ground defense was impossible. In the Japanese phase of the Second World War, air power was decisive, and rather than face destruction from the B-29's the Japanese capitulated.

The extent to which the strength of air power mounted is shown succinctly by the bomb-tonnage figures:

	Year	*Tons by* USAAF
The Second World War in Europe	1942	6,123
	1943	154,117
	1944	938,952
The Second World War in the Pacific	1942	4,080
	1943	44,683
	1944	147,026
	1945	1,051,714‡
	1946	3,167,316‡

* When the studies of the U. S. Strategic Bombing Survey in Japan are completed, it will be possible to use up-to-date figures.

† See Dr. Oppenheimer's estimate, page 25.

‡ These figures, which do not include atomic bombs, include projected tonnages for the latter part of 1945 and 1946.

The harnessing of atomic energy and its application at the climax of the Pacific War have tended to overshadow an important point. Even before one of our B-29's dropped its atomic bomb on Hiroshima, Japan's military situation was hopeless. The projected tonnages quoted could have forced capitulation no later than 1946. Without attempting to depreciate the appalling and far-reaching results of the atomic bomb, we have good reason to believe that its use primarily provided a way out for the Japanese government. The fact is that the Japanese could not have held out long, because they had lost control of the air. They could offer effective opposition neither to our bombardment nor to our mining by air and so could not prevent the destruction of their cities and industries and the blockade of their shipping.

recaptured; but during the time of its use, they operated at a good margin of profit and did damage and forced countermeasures that cost the Allies many times as much effort as the campaign cost the Germans.

Our increase in offensive air strength came about from three factors, all of which must be borne in mind in planning for the future.

1. *Increased size.* Both aircraft production and the Army Air Forces expanded enormously. These facts are so well known and appreciated that they need not be discussed here.

2. *Improved quality and efficiency of the instruments of air power.* These were planes, electronic equipment of all sorts, and greatest of all, explosives, including the atomic bomb. Improved quality is not obtained overnight; it comes about only

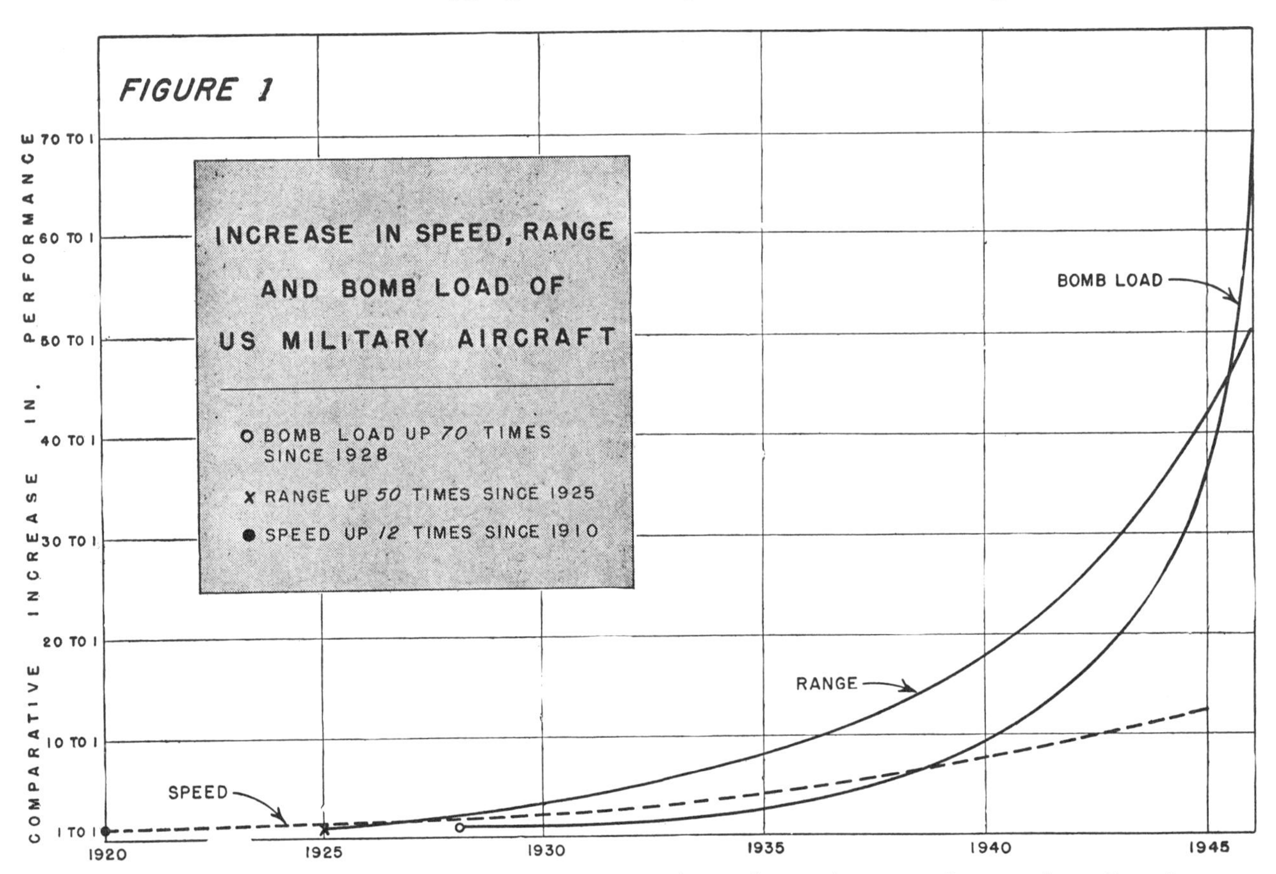

Air power was advancing in the hands of the Germans also, although along different lines. Prevented by Allied mastery of the air from using piloted bombers effectively against England, they turned to alternative air power weapons and developed V-1, the jet-propelled, pilotless, expendable bomber, and V-2, the logical extension of jet propulsion to stratospheric conditions. They were forced to abandon V-1 when the French coast was

through persistent and extensive effort in research and development. As a consequence of the improvement in quality, the B-29's delivered bombs at three times the range for half the cost per ton compared to heavy bombers in Europe. Figure 1 shows progress in design in a more general way. The curves corresponding to speed, range, and bomb load for aircraft in use at the dates indicated show a steady increase during peacetime that was greatly accelerated during the war.* This acceleration reflects increased speed of getting new models into production, as well as intensified research and development. If our intensified research program had preceded the war, new types of combat aircraft would have saved many American lives. The lesson from these facts is that research must continue during peacetime and cannot be slighted until the war clouds are seen to be gathering. This will be even more true if what we face is a war with atomic weapons.

3. *Increased efficiency in using the weapons of air power.* Improvements in our training program and in the know-how of using the weapons in combat produced great increases in combat effectiveness. The bomb density on the target for visual bombing (Fig. 2) nearly doubled between 1943 and 1945. This was due not only to increasing skill but also to increased mastery of the air as German fighters were thinned down. The amount of flying each aircraft did was more than doubled between 1943 and 1945 (Fig. 3). This increase was due to improvements all along the line, including maintenance, organization of large missions, and decreasing enemy resistance.

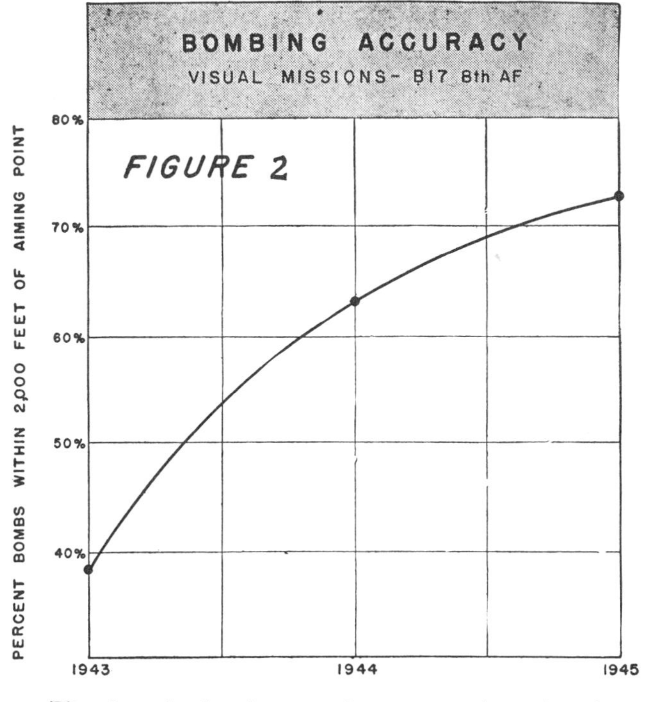

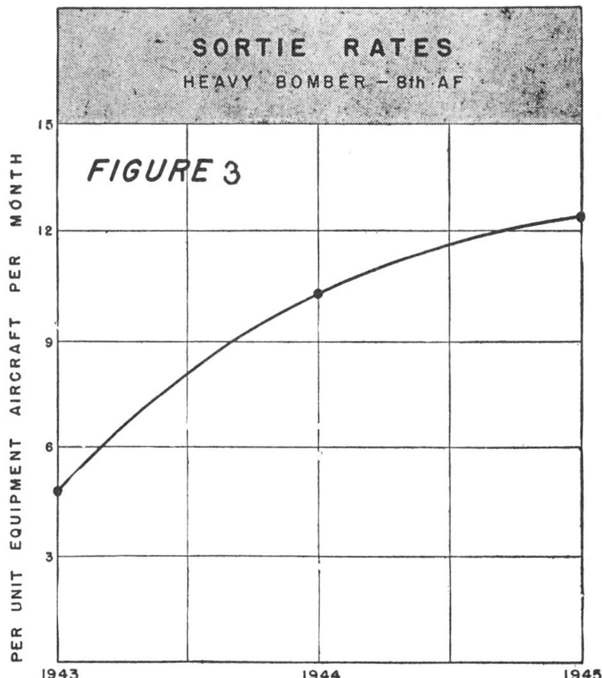

The Future of Air Power. Before the end of the Second World War, one B-29 dropping an atomic bomb caused as much damage as 300 planes would have done before. A consequence of this fact is that the Twentieth Air Force, using atomic bombs, could in one day's raid destroy more of Japan's industry than was actually done in the entire B-29 campaign. This damage amounted to over 42 per cent of the urban industrial area of the sixty-eight cities attacked. These cities had a total population over 21 million—almost exactly equal to our 12 largest American cities. This damage, I repeat, would have required less than one day's effort by a force the size of the Twentieth Air Force. The cost of the atomic bombs on a mass-production basis, according to Dr. Oppenheimer's unofficial estimate, would be less than 200 million dollars—a very modest sum in wartime. Continuing this program for a few days or weeks would obliterate all the industrial centers in the Japanese Empire. A similar fate would have enveloped England had the Germans obtained atomic explosives for their V weapons.

We have now flown a B-29 nonstop for a distance of over 8,000 miles. Ranges of 10,000 miles for one-way trips with newer types of aircraft appear to be a possibility of the near future. The strategic significance of these facts is at once evident when one draws the radii on a map. It means that with relatively unimportant exceptions all of the centers of civilization in the northern hemisphere are within reach of destruction at the hands of any major nation in that hemisphere.

* Actually the intensification of research and development after Pearl Harbor, while improving the characteristics of previously designed types, did not lead to the introduction of any new types in combat. All our combat types were designed before we entered the war.

Looking only a little farther to the future, we must consider developments of the V-2 rocket. By designing a rocket consisting mostly of fuel, a speed of 3,400 miles per hour and a range of 200 miles was obtained in operations against England. This rocket weighed 14 tons and delivered only 1 ton of explosive. The Germans had under design a longer range rocket, using a pick-a-back principle. A large rocket was to carry a smaller one up to a speed of 2,500 miles per hour. At this point the smaller one was to take off on its own, attaining a speed of 5,800 miles per hour, which would carry it 500 miles. The combined rocket would weigh 110 tons and deliver 1 ton of explosive. Designs incorporating winged rockets predicted an increase of range to 300 miles for V-2 by finishing the trajectory in a glide and an increase to 3,000 miles for rockets of the 100-ton class. This phenomenal increase in range was based on a trajectory in which the rocket would bounce out of the lower atmosphere into the stratosphere in a succession of jumps (like a stone skipping on water) and end in a glide. Extensions of these techniques to rockets of more than two stages permit increasing the range indefinitely with a progressive decrease in per cent of payload. However, the achievement of range alone is of little strategic value unless it leads to effective and economical destruction of specific targets.

The problem of transporting explosive—or perhaps mail, materials, or even persons in peacetime commerce—to an assigned destination cannot be satisfactorily solved simply by aiming the rocket initially like a projectile from a gun. Instead, it is necessary to correct the aim and guide the rocket during the stratospheric portion of its flight. In the case of V-2, the guiding and control was restricted to the first 66 seconds of flight during which the rocket climbed 12 miles. The average error on a 200-mile shot was four miles—satisfactory for bombarding large cities such as London. But the same degree of control on a 3,000-mile shot would lead to an average error of 60 miles—in other words, only one in 600 rockets would hit a city the size of Washington.

Obtaining satisfactory accuracy first requires finding the correction in path necessary to make the rocket hit the target, and second, altering the rocket's course to accord. Methods now in use based on radar principles satisfy the first requirement with one mile accuracy at 100 miles and two miles at 600 miles from the control stations. No basic principles are known which will prevent achieving these or greater accuracies at much larger ranges. The problem of altering the rocket's flight will probably involve deflecting the wingless types by rocket jets during the stratospheric portion of the flight and using conventional control surfaces for the gliding and skipping types. The most pertinent comment about rockets that can be made now is that research and development will improve their characteristics much as aircraft characteristics have been improved.

Very long-range rockets were not a serious threat before the atomic bomb because of the high ratio of the weight—and consequently the cost—of the vehicle compared to the explosive load. Even a two-stage rocket would barely break even economically—the cost of destroying a square mile by this means being comparable with, if not larger than, the damage done to the enemy. With an atomic warhead, however, the cost of the carrier is not excessive and even the expense of the most elaborate guiding and control equipment would not make the total product inefficient, for the destructive effect would exceed the total cost by a large factor.

The problem of defense against V-2 was unsolved during the war, and, although interception by ground-launched rockets was theoretically possible, the practical difficulties would take years to overcome. The pick-a-back rocket and other offensive improvements could, on the other hand, be developed in much less time.

Jet propulsion is another signpost to the future. The limitations imposed by propellers will be eliminated, and ceilings will be lifted on the speed and altitude of conventional planes and their future developments. The most serious limitations in aircraft performance come from the phenomenon of compressibility, which occurs when the velocity of the aircraft approaches that of sound. This is in part responsible for the fact that the rise in the speed curve (Fig. 1) is less sensational than the rise in the other curves shown. Now that jet-propelled aircraft have become a reality, the speed of sound no longer appears as an unsurmountable limit, and aircraft with supersonic speeds flying at stratospheric altitudes are thought to be within reach of a few years' development. Pilotless weapons based on such aircraft have potentialities at least comparable with the developments of V-2.

Future Requirements. Against this future of increasing range, speed, and destructiveness of the weapons of air power, adequate protection by pure defense seems unlikely. Our defense can only be a counteroffensive; we must be prepared to give as good as we take or better. Should we ever find ourselves facing an aggressor who could destroy our industrial machine without having his destroyed in turn, our defeat would be assured. Thus our first defense is the ability to retaliate even after receiving the hardest blow the enemy can deliver. This means weapons in adequate numbers strategically distributed so that no enemy is better situated to strike our industry than we are to strike his.

The picture of defense by counteroffensive is a gloomy one. A far better protection from atomic weapons lies in developing controls and safeguards that are strong enough to prevent their use on all sides, for that offers the only hope for preserving the values of our civilization. Still, it is my duty here to trace what must be the air-force policy of the United States in the absence of such controls—and a grim duty it is for any peace-loving American.

Our countermeasures to anticipate and block an aggressor's blows, so long as such blows are possible, must be developed to the utmost. Since, in the near future, we expect that offensive air power will outstrip defense and become adequate to accomplish almost any degree of destruction, the nation that first develops a means of protecting itself will be the first able to initiate an atomic war without simultaneously bringing equal destruction on itself. Thus we must make sure that no potential aggressor outdistances us in his defense developments. And this means strategic bases for warning, detection, and interception of bombers for the near future and later for pilotless weapons.

One passive defense of great importance consists of dispersing or burying below ground essential war industry. At present, with our industry intact after the war, we are the least dispersed of all major nations. In other nations, dispersal either has already come about as a result of war or will be incorporated in rebuilding destroyed industry. It is worth while to distinguish between two kinds of dispersal. One kind is required to prevent fatal disorganization by a relatively small number of accurately placed atomic mines. This would have to be achieved well in advance of open hostility. The other is sufficient dispersal to permit us to carry the offensive back to any nation daring to make an aggressive attack. In a world in which atomic weapons are available, the most threatening program that a nation could undertake would be one of *general* dispersal and fortification. Should such activities start, the world would see the greatest digging race of all time—and the greatest war.

However, not only must we prepare to fight an atomic war, but—strange though it may seem at first—we must also prepare to fight a war in which no city may feel the blast of an atomic bomb.

The possibility of a future war without atomic destruction of cities is not based on the belief that the enormous devastation expected of atomic weapons will lead to such abhorrence of them that they will not be used. In the past no effective weapon of war has remained long unused, and in fact the atomic bomb has already brought destruction on Japan.

But in the past destruction by ground forces has been largely in one direction, and whichever side had mastery could destroy without suffering retaliation. The anticipated cheapness of destruction with atomic explosives, coupled with the fact that their use with air-power weapons such as rockets prevents adequate defense, means that in the future the aggressor who destroys his enemy's cities may expect his own destroyed in turn. Neither side may care to fire the first atomic shot and thus bring destruction to its own cities and population as well as to the enemy's. In addition, a nation planning aggressive war in order to enrich itself by mastering its opponent's industrial and economic wealth may withhold some atomic bombing to avoid obliterating its contemplated loot.

There is historic precedent for withholding destruction in wars. The case of gas in Europe is an example. It was used by neither side and the reason in Germany's case was in part, at least, that she did not care to have gas used on her. Other instances of non-destruction, although of a somewhat different type, are furnished by the open cities of Paris and Rome. A related case is that of Switzerland. While not an Axis nation, Switzerland was safe from Axis attack, partly because of her ability to retaliate by destroying her strategic railroad tunnels which would have greatly impeded the flow of materials such as coal between Italy and Germany.

Now the arguments given above are not intended to comfort us with the thought that, if all nations had atomic weapons, no nation would use them for fear of retaliation. All they show is that there is a *possibility* of stalemate with respect to destruction of cities by atomic bombs. Preparedness, although it must be built around atomic weapons, therefore *cannot* be built around atomic weapons *alone*. Proper account must be taken of the other forces of land, sea, and air. Still another reason for maintaining a balanced armed-forces program is that, after major atomic destruction is accomplished, a war might not be settled, and a conflict using atomic weapons only in part might well continue.

It is worth while pointing out that biological warfare, consisting of the spreading of disease, could occupy a position similar to atomic warfare and that the same arguments would apply to it.

For the air force to carry out its mission in the future, we must have

1. An up-to-the-minute air force with trained personnel and with all the latest and most efficient air-power weapons, quite apart from atomic weapons, quite apart from atomic explosives. Our counteroffensive must be made by a force in being —not a force which has to be mobilized in weeks or even days.
2. An extensive and efficient intelligence service to make available a maximum of warning of worsening relations or impending attack.
3. Research and development adequate to make our equipment the foremost in the world. In this connection great caution must be observed that we do not defeat the very object we are trying to achieve by trammeling our scientists with security regulations and thus rendering them less able to cope with future emergencies.
4. Strong industry capable of rapid expansion for war production.
5. An integrated national-defense organization geared to the new concepts of total war.
6. Strategic bases.

In the foregoing analysis, I have taken the point of view which I must take—that adequate international controls and safeguards have not as yet been established, and that the air forces must therefore be prepared to protect the nation in a future war. How terrible that war would be is indicated by the factual appraisal of what atomic explosives mean in comparison with the most deadly materials men have devised before them. I have mentioned the possibilities that might lead aggressors to refrain from using atomic bombs in a hypothetical future war. Let me say again that, although they are far less economical or efficient, conventional bombs as they will be developed before that war can come will go far enough toward wrecking the world. Thus, whether we recognize that atomic bombs will rain upon us or cling to the faint hope that only standard high explosives will batter our cities, we must realize that the time is at hand for the peoples of the world to admit that their warring power is too great to be allowed to continue. Through international collaboration we must make an end to all wars for good and all.

Chapter 7

LOUIS N. RIDENOUR *had charge of the development of several types of radar at the Radiation Laboratory of the Massachusetts Institute of Technology. During 1944 he served in Europe as radar adviser to General Spaatz. He returns this year to the University of Pennsylvania as professor of physics.*

There Is No Defense

by LOUIS N. RIDENOUR

THE subject of this chapter is the possibility of active defense against an attack in which atomic bombs are delivered by means such as those described in the last chapter. As a means of getting into this, I should like to project into the future what we learned regarding active air defense in the war just past.

We must first consider whether it is necessary to shoot bombs down at all in order to render them ineffective. Is there not some specific countermeasure which would keep the bomb from exploding, or cause it to explode harmlessly at a great distance from its target? Many authorities and semiauthorities, including the House Naval Affairs Committee, have made public statements which encourage this hope.

All such statements are very dangerous because of the mistaken complacency they can engender. There is no such thing as a specific countermeasure against an atomic explosive or an atomic bomb. For that matter, there is no specific countermeasure against an old-fashioned chemical explosive such as TNT or black powder. All explosives, including atomic explosives, go off when set off by their detonating mechanisms, which are always made rather simple and tamperproof. This has been made abundantly clear by wise and responsible authorities, but it cannot be said too often.

Many of the misleading statements about hopeful countermeasures have been made by honest men who have no active desire to mislead. Themselves misled, they understand the problem incompletely. Their reasoning goes somewhat like this: It is assumed that the detonator of the bomb, or the steering mechanism of the vehicle which carries it, or some other vital device connected with the bomb, takes a certain form. A means is then devised for interfering with the operation of this device. The inventor of the countermeasure forgets entirely that if the enemy does not make use of exactly the form of device imagined, his idea is useless.

A specific example may make this more concrete. As already announced, the atomic bombs dropped on Japan were caused to explode about 1,500 feet in the air. The fuze may have been connected either to a barometric switch set to close at an air pressure corresponding to the desired altitude, or to a radar altimeter. The latter device measures altitude by sending out radio waves and measuring the time lapse before they bounce back from the ground or sea beneath. If a hypothetical enemy used against us bombs equipped with a radar altimeter, we *might* be able to send out interfering radio signals which would set off the bomb at a higher altitude than its designers intended.

We could scarcely expect to make the bomb go off while still in the bomb bay of its carrier, or at a very much greater altitude than that for which the altimeter was set, for the greatest attention is paid by ordnance experts to arming devices designed to prevent this. Such devices keep bombs and shells in a safe condition, with the fuze mechanism entirely inoperative, until a considerable time after they have been dropped from the carrying aircraft or fired from their guns. Mainly intended to make bombs and shells safe to handle and not too dangerous to the gunners or air crew who must deliver them, arming devices also serve to defeat many schemes to produce ineffective premature explosions.

More important than the fact that a countermeasure directed against the radar altimeter could at best be only partly effective—since the bomb still would explode at an altitude at which damage would result—is the strong possibility that the enemy might be so unobliging that our countermeasure would not work at all. If he used a barometric altimeter instead of a radar altimeter, there is nothing that we could do about it.

Simple and direct habits of thought lead to weapon design giving little leverage for countermeasures. The V-1 flying bomb was a good example of a weapon that did not lend itself to any sort of countermeasure except direct interception, and V-2 was an even better one. The V-1 was steered after launching by a magnetic compass. It was kept at the desired altitude during its flight by a barometric altimeter. It had an automatic pilot and a means for causing it to execute a gentle turn after launching, if desired, but the former was a simple and rugged gyro device, and the turn mechanism ran by clockwork. Fragments of radio equipment were found in some of the V-1 wreckage and gave rise to hopes that the Germans were using radio control that could be interfered with by British countermeasures. This was not the case. The radio parts came from simple transmitters, which the Germans put into a small fraction of the bombs so that radio-direction-finding stations along the Channel coast could plot their paths and thus keep track of wind conditions over England.

People talked of a great magnetic coil overlying the Downs, which would produce enough change in the earth's magnetic field to affect the magnetic compass of the V-1 and thus deflect the missile from its objective. Calculations showed that no sufficiently strong field could be produced, and anyhow the unobliging Germans had fixed V-1 up so that even a violent deflection of the compass would put the vehicle into no more than the gentlest of turns. There was nothing to do but to shoot the V-1's down in the old-fashioned way, with fighters and antiaircraft guns.

The pained disappointment this caused the countermeasures experts was nothing to that caused by the properties of V-2. Our first detailed knowledge of V-2 came from fragments of a trial shot that fell in Sweden. The rather well-preserved wreckage yielded a good deal of radio equipment. It was hoped that this meant radio control of V-2, for the fighter planes and antiaircraft that had been used against V-1 were about as useful against V-2 as trained falcons and a bow and arrow. Radio control, which might admit of jamming and countermeasures, seemed the only defense possibility.

Presently V-2's began to land in London. Countermeasures men spent a good deal of hazardous and unpleasant flying time in the vicinity of the launching sites, hoping to pick up, identify and study radio control signals as a preliminary to jamming them. But there was no radio control; the Germans were simply firing V-2's at the target and hoping for the best. There was no countermeasure short of winning the war, which, luckily, was done. But there would not have been time for that if the V-weapons had had atomic war heads.

The public imagination is readily caught by a comic-strip sort of ray or stream of energy that will destroy a missile at a substantial distance from its destination. A simple calculation shows that the destruction of a single missile by such means would drain all the power from the section of the country in which the installation was located. Such grave practical questions of the feasibility of dealing with a single projectile by this means are entirely overshadowed by the obvious impossibility of dealing thus with a mass raid, which we must expect a possible attacker to deliver.

In half a century of dealing with chemical explosives, no one has yet discovered how to detonate them by a ray or by any other setup that works without wires at a distance. It continues to be overwhelmingly unlikely that such a discovery will be made in behalf of atomic explosives.

In the light of everything we know about the atomic bomb, there is no such thing as a specific countermeasure.

We are faced, then, with the problem of mounting an active defense against the carriers of atomic bombs. We must detect them and follow their paths by radar, plot their future courses with predictors, and engage them with countermissiles before they have approached near enough to their target to be dangerous. What are our chances of preparing an active defense affording protection against an atomic bomb attack which may come without warning?

Our chances of doing this are vanishingly small. To understand why, we shall have to look into the character of the defenses, so far as we are able to predict them on the basis of our experience in the past war and our knowledge of later technical developments that did not come soon enough to see service in the war. The problem separates itself into four main parts—detection, identification, course prediction, and interception.

Detection will certainly be accomplished by radar, for there is no other tool that can be relied on to operate under all conditions of visibility at all times. The best of the search radar with which we finished the war could see a single heavy bomber out to a distance of 200 miles or so and covered a region of space extending up to an altitude of 40,000 feet. Radar search equipment for atomic defense will have to provide coverage over the entire upper hemisphere, for high-angle projectiles derived from the V-2 rocket must be detected. By the same token, it will not be enough to install radar simply around the coasts and borders of our country. We shall have to defend St. Louis by near-by installations, for the arching trajectory of a long-range rocket could bring it across the coast at an altitude of hundreds of miles—where it would be extremely difficult, if not impossible, to pick up—and guide it down on its inland target undetected, unless there were inland radar installations.

Present techniques make a radar detection range of about 200 miles for a missile the size of V-2 not impossible, though the supersonic streamlining of this vehicle makes it difficult to see by radar. Proper coverage would probably demand that about 250 long-range search radar installations be operated on a twenty-four-hour basis to provide adequate detection facilities. Each installation would need perhaps five separate radars, to provide off-the-air maintenance time and to give adequate hemispheric coverage. The crew for each installation would be about 200 men; the equipment cost of a single installation would be about 1½ million dollars. We should thus have to spend upward of 375 million dollars to make these installations and employ some 50,000 men in operating them. If we did, we should be able to detect aircraft and air-borne missiles with considerable reliability.

Indeed, our principal difficulty would then be a sort of embarrassment of riches; for radar detects friend and foe alike. In the war just past, a special radar beacon, which gave a coded reply to a specific challenge, was originally meant to be carried in every ship and aircraft of the United Nations. This equipment was called IFF (Identification of Friend and Foe). It was an abject operational failure. In the European Theater it gave so many difficulties that its use was entirely abandoned after D-day in Normandy, except for small numbers of aircraft on special missions. In the Pacific, where aircraft densities were lower, its use was continued; but at the times when it was most sorely needed, IFF failed to serve its intended purpose. The main troubles with wartime IFF arose under conditions of heavy traffic—when there were many aircraft in the view of a single radar set.

For our future defense we shall have to supplement our radar search and warning chain with an IFF system which will work even in higher traffic densities, will be quick and reliable in use, and that will inherently not be subject to compromise by a possible enemy. This last requirement presents the utmost difficulty, since the system must be so universally used that every potential enemy will have full opportunity to become familiar with its principles, its detailed design, and its prescribed use. Despite all this, the IFF must be capable of being employed in such a fashion that friendly craft can be distinguished from enemy vehicles.

The restriction of air traffic to prescribed lanes, and the regulation of traffic within those lanes, will undoubtedly be of assistance to the radar stations in keeping track of air activity. But even with all the aids to identification that ingenuity is able to provide, future air traffic will probably be so dense that the identification problem will be a major one.

Even so, let us conclude that an enormous national investment in radar and IFF facilities will give us some chance of detecting atomic missiles while they are still one or two hundred miles from their target, and of recognizing that they are not friendly. We are now faced with the problem of destroying them before they have come dangerously close to the target. In the air defense of the last war this was done either by means of piloted fighter aircraft or by means of artillery shells. In either case, such interception of the enemy craft was preceded by a course prediction that based the expected future path of the craft on a knowledge of its past course. Such course prediction was performed for antiaircraft artillery by an electrical computer at the gun battery. It was performed for the fighter either by the pilot himself, if visibility permitted, or by a controller on the ground who watched the radar plots of both enemy craft and friendly fighter.

Since the device used for course prediction depends fundamentally on the device used for interception, it will be convenient to consider the two problems of course prediction and interception together. What sort of vehicle will be used as a defensive interceptor? It seems certain that it will be neither the piloted fighter nor the conventional antiaircraft artillery shell, for both of these were already obsolete by the end of the past war.

The design of the pilot, and not of the fighter plane, made the piloted fighter obsolete. As aircraft speeds went up, maneuverability—necessary to meet evasive action by the target or for last-second adjustments in aim—decreased, for the acceleration permissible in a turn is restricted by the pilot's black-out at high accelerations. In turns so sharp that the pilot is subjected to a centrifugal force larger than $6g$—six times the force of gravity at the earth's surface—his vision dims out, and he may even lose consciousness. Turns much sharper than this cause bursting of blood vessels, detachment of internal organs, and death. And, of course, the geography covered in the course of a simple maneuver of a high-speed aircraft is immense.

A pilotless vehicle can be built to make turns of arbitrary sharpness at any speed. It is necessary only to make the structure strong enough so that the acceleration does not tear it apart and to make provision for plenty of control force. There are at speeds near that of sound certain aerodynamic limitations that are neglected here because these difficulties arise even in the problem of level flight at very high speeds. What is important to our argument is that limitations put by the design of the human body on the maneuverability of a high-speed interceptor are so severe that our expectation of high speed in attacking vehicles will demand pilotless interceptors.

The simplest type of pilotless interceptor is the artillery shell, but this is now obsolete for the same reason that the piloted plane is obsolete—the speeds of the target vehicles have gone too high. The piloted fighter cannot make sharp turns at high speed; the artillery shell cannot make turns at all. The shell's effectiveness depends on our ability to get it as rapidly as possible from gun barrel to target, so that our prediction of where the target will be when the shell arrives is not outdated by unanticipated maneuvers of the target during the time of flight of the shell.

The remarkable electrical predictors used by our antiaircraft to compute the future position of a target on the basis of its past course and speed were and are subject to several serious limitations. First, they assume flight in a straight line. Second, no computer, however ingenious, can allow for a target maneuver that is undertaken while the shell is in flight.

But a target going 500 miles per hour, if it should commence a $6g$ turn at the moment a shell was fired (acting perhaps in response to the gun flash), could be $14\frac{1}{2}$ miles away from its predicted position and going the other way if the time of shell flight were only 13 seconds. A faster target, or one capable of evasive action at higher values of acceleration, could correspondingly escape aimed fire more readily. The solution to this difficulty clearly lies in cutting down the time of the shell's flight, which depends on the distance to the target and on the speed of the shell. But even with guns made of the strongest metals we now have, muzzle velocities are not more than a few thousand feet per second. And the danger range of the atomic bomb is so great that an antiaircraft battery *must* engage and destroy its target at long range if it is to afford any protection at all to the vital area it is guarding and if the antiaircraft battery is not to destroy itself with its first successful shot.

In addition to the fact that evasive action and increasing target speeds limit the effectiveness of antiaircraft fire, we must consider the inherent inaccurary of this gunfire even under the best conditions. Errors may arise because the target is not being followed precisely by the fire control radar or because the target has not been followed for a long enough time to make possible an accurate prediction of its future course or because the computer makes errors or approximations in its prediction—or for many other reasons. It is sufficient for our purpose to point out that in the Second World War antiaircraft fire was operating just on the edge of the region where aimed fire could not bring hits without a large portion of luck, and V-2 carried us well into that region. There will be no use in firing even the best imaginable antiaircraft artillery at the missiles that might be the first harbingers of a new atomic war.

Faced with the obsolescence of the piloted defensive fighter and of antiaircraft artillery, it seems most likely that our military designers will develop a vehicle which is launched at high speed in accordance with a predicted target position (like a shell), and that has motive power, flight controls, and a target-seeking mechanism (like a piloted fighter). All this can be done. We can conceive of a missile that might be fired from a sort of gun, propelled by a rocket charge or other means during later flight, and caused to home to its target by radar or other target-seeking means. No doubt a great deal of thinking and probably some work is going into the design of such missiles right now.

The proximity fuze, which explodes an antiaircraft shell when it comes close enough to its target to be able to destroy it, is already in hand. Perhaps an atomic warhead would also be used on new defensive missiles to increase their radius of effectiveness.

This completes the picture of our possible active defenses. We should have to spend a lot of money and effort designing and installing a radar chain for search and warning, but let's believe that the resulting equipment would search and warn. We should have to be very clever in our design of identification equipment, but let's believe that proper traffic control measures would enable us to tell peaceful air traffic from the missiles of a possible enemy. We should have to spend a great deal of ingenuity and effort in the design of a homing antiaircraft rocket missile. Let us believe that we could do all these things. There is much here that we cannot now do and no one can say with certainty that we shall eventually be able to do all that we would have to do. But *if* we could, would this chapter still have to be headed There Is No Defense?

This is a perfectly good question on the basis of what has so far been said, and it has a perfectly good answer. The answer is that a war, or even a new phase of a war already in progress, *always* starts with a Pearl Harbor kind of attack. In a war involving old-fashioned explosives, we have called what happened while we were getting our guard up a disaster, but we survived it and went on to fight. In an atomic war the first attack, no matter how well prepared for it we may be, will *really* be a disaster. It is quite likely to end the war, if we have a practical, ingenious, and determined enemy.

It is possible to go on and say that no defense against any form of air attack has ever achieved 100 per cent success at any stage of experience, but this is somewhat redundant if we are convinced by an examination of the past that a first attack always succeeds in achieving surprise.

Pearl Harbor itself, besides giving its name to the phenomenon of costly unpreparedness, is a very good case in point. Even the primitive radar equipment installed for the defense of the island succeeded in giving warning that would have been adequate to prepare a vigorous defense had the defenders been alert. No one could have wanted more eagerly than the Army and Navy commanders then at Pearl Harbor to be alert, to avoid a debacle, since it would—and did—blight in a morning the careers of a lifetime. These men were simply victims of the principle that the defense can never be ceaselessly alert everywhere and that the offense can always select the place and time to strike.

Even when there is warning the situation remains about the same. Early in the winter of 1943–44, the British were perfectly well aware of German preparations to attack London with V-1. They had good information on the construction and characteristics of V-1, and they studied carefully the various possibilities. Yet when the flying bombs started to come in July, 1944, the British were far from ready to meet them. In the first days of the attack more than 35 per cent of the V-1's launched by the Germans struck London. Eight weeks later, with no more gun batteries available, only nine per cent of the V-1's launched were reaching London. In an atomic war even the nine per cent figure would be fatal to London in less than a single day, but the 35 per cent figure is the one to ponder.

By the time we deployed our antiaircraft defenses to protect Antwerp from V-1 attack, we knew all about the buzz-bomb. We knew the best disposition for batteries and the best plans for their operation. The gunners had had experience in the defense of London, and knew what they had to face. Yet the initial efficiency of the Antwerp defense was only 57 per cent. During the next two weeks it rose to over 90 per cent.

All through the war just past there were evidences of this same phenomenon of initial defense inefficiency. When the Germans began their daylight air attacks on England in 1940, the RAF had to organize its defense, which was rather ineffective at first. The RAF defense soon became so effective that the Germans shifted to night bombing, but not before considerable damage had been done to England. The technique of ground control of night fighter interception was worked out quickly by the RAF, the German losses in night bombing began to mount, and presently these losses became so prohibitive that the Luftwaffe gave up—but again considerable damage was done to England. In an atomic war either of these campaigns of the Luftwaffe, quickly opposed and mastered by the RAF as each was, would have wiped out England.

We can sum up the situation regarding active defense against airborne atomic bombs in the following terms:

1. There is no such thing as a specific countermeasure that will prevent the explosion of an atomic bomb or will explode such a bomb while it is still a great distance from its target.

2. A radar detection net which could in principle provide warning of the approach of air-borne atomic bombs is technically feasible. It would represent a tremendous national investment and a continuing drain on manpower.

3. The chief difficulty connected with radar detection of missiles directed at us in a future atomic war would be that of separating the radar signals produced by such objects from those caused by friendly and normal air traffic. This calls for the development of an identification system of unparalleled effectiveness and subtlety, together with the imposition of stringent rules of traffic control.

4. Interception of missiles used for the delivery of atomic bombs, which we can expect to travel with supersonic speeds, will demand a totally new development. Interception devices which are themselves capable of these speeds and which can seek out and home on their targets will be necessary. Such interception devices may be developed but will require a very considerable effort and tremendous expense. Piloted fighter aircraft and conventional antiaircraft artillery will be entirely useless.

5. Regardless of our state of preparedness for an atomic bomb attack, it is likely that the initial effectiveness of our defenses will be small. The defense must be eternally alert everywhere, for the offense strikes at a chosen time and place. Personnel training in time of peace is no substitute for wartime training when the chips are down. We never anticipate the exact character of the actual attack nor the defense strategy best adapted to meet it.

6. If an atomic-bomb attack does not devastate its target in the early days of low defense efficiency, the maximum efficiency that can be expected of the active defenses when the defenders are experienced is around 90 per cent. With attacks carried out by piloted bombers, such efficiency is murderous and would be an effective defense; with attacks carried out using conventional chemical explosives, even with pilotless missiles, such efficiency would raise a serious question of the economic wisdom of the enemy continuing the attacks. With atomic explosives, 90 per cent efficiency is definitely too low to afford adequate protection. The destructive effect of the 10 per cent of incoming missiles penetrating the defenses would be great enough to wipe out the target, even though the attack were carried out on a modest scale.

7. There is no defense.

E. U. CONDON, *born in Alamogordo, New Mexico, was associate director of the Westinghouse Research Laboratory from 1933 to 1945 and is now head of the National Bureau of Standards, as well as adviser to the Senate Atomic Committee. He served on various uranium committees from 1941 to 1943.*

Chapter 8

The New Technique of Private War

by E. U. CONDON

LOWER New York Harbor shook that day in 1916. Some freight cars and a barge loaded with TNT and picric acid on their way to the armies of the Czar of all the Russias had blown up. The Black Tom explosion was typical of successful ventures in sabotage, a prototype of this stealthy tactic of total war. Agents of the German government had detonated the shipment of high explosive by secreting small time bombs in the loaded cars. The ill-equipped Russian army had suffered a costly defeat at the hands of a few careful and determined men.

In Rjukan, Norway, the cheapest hydroelectric power in the world encourages the production of hydrogen by electrolysis of water, and the residue from the electrolytic cells of the great Rjukan plant is incidentally enriched in deuterium. Further treatment of these residues yields pure heavy water, which can be useful in the production of plutonium. When the Nazis undertook such further treatment of the residues, the British—properly apprehensive of the atomic bomb in Axis hands—took counsel with the Norwegian underground. His Majesty's Government armed and encouraged the saboteurs in their spectacular attacks on the heavy-water plant. Its structure and purpose were so special, and the necessities of its location so restrictive, that the crippling of this one plant would seriously reduce the German facilities for manufacturing heavy water, hence plutonium and bombs.

These two examples exhibit the major principle of all important war-time sabotage. The saboteur cannot have mighty engines or tons of explosive at his disposal; he moves by stealth and carries his destructive means in his pockets or on his back. In the days before the atomic bomb this left him two choices. He could destroy a small but important target with the explosives he carried himself, or he could touch off with this tiny charge the energy stored in the munitions of his enemy. Rjukan or Black Tom—the indispensable small thing or the great concentration of unstable explosive. Both were vulnerable to the old-style saboteur.

Because of this, both were surrounded with safeguards. The small thing, being small, can be specially and heavily guarded. The sentry walks the bridge; the President is accompanied by his guards; the cars entering Oak Ridge are searched. The load of explosives can be guarded too, and the munitions plant or dump is isolated both from the abodes of men and from other concentrations of explosively unstable chemicals. When a shipload or a trainload of explosives comes to a port, it goes by circuitous routes to a lonely spot where its explosion will do little harm. The disaster of Port Chicago, where a shipload of naval explosives detonated for reasons unknown, typifies the unexpected explosion. Men working there died by the hundreds, but the site was so remote that there was not a single civilian casualty.

In the age of atomic explosives the special agent has not been freed from the traditional restriction of his profession—his physical means must still be small. But no longer does this connote small destruction. No longer must significant damage be done by painfully gaining close access to a vulnerable target. No more must he study the habits of the pacing guard and slip past to put the few pounds of TNT and primacord against the generator. No longer need Guy Fawkes put the gunpowder directly under Parliament. The atomic bomb of modest size that the agent assembles in his hideaway will, when it goes off, take with it every structure within a mile. Within the volume of a small watermelon is stored the energy of more than 20,000 tons of old fashioned high explosive. The saboteur can carry on his person more destruction than the Eighth Air Force could bring to

Germany by ten raids at maximum effort; ten raids in which 200 heavy bombers and 2,000 airmen would be lost.

We must, therefore, no longer expect the special agent to be special. In a future war and perhaps even in the months of tense suspicion that may precede it, the activities of the saboteur will be important. Against him the locked door and the armed guard no longer can prevail. A target, to be safe, must be surrounded by a sanitary area of at least a mile in radius, all known to contain no suspicious man or thing. Any house can be as dangerous to its surroundings as the greatest of powder magazines. Twenty-thousand tons of TNT can be kept under the counter of a candy store.

This does not overstate the case. To be sure, the little lump of atomic explosive must be put together with other mechanisms to make an atomic bomb. Some chemical explosive must be used, with a massive surrounding bubble called a "tamper." Our government is chary of the details, but we know that the resulting bomb will fit the bomb bay of a B-29, and we can be sure that the structure can be made for a total weight not far from a ton. It can be packaged in the shape and appearance of a filing cabinet or an upright piano.

Can it be detected at a distance by its radiations? Robert Oppenheimer, asked in Senate hearings whether there were not some scientific instrument that would enable the detection of the exact box in a Washington basement that might contain an atomic bomb, answered: "Yes, there is such an instrument. It is a screwdriver, with which the investigator could painstakingly open case after case until he found the bomb." Oppenheimer was not joking. Small amounts of radiation are emitted by uranium-235 and plutonium, but the heavy metallic tamper used to make the bomb efficient serves well to absorb this already weak radiation. Neutrons emitted by the atomic explosive are especially penetrating, but the bomb is so constructed that before detonation there are few neutrons present, and the entire design is devoted to preventing their escape. Encasing the bomb in a wooden box would screen it from inspection no more remote than the next room.

We must accept the fact that in any room where a file case can be stored, in any district of a great city, near any key building or installation, a determined effort can secrete a bomb capable of killing a hundred thousand people and laying waste every ordinary structure within a mile. And we cannot detect this bomb except by stumbling over it, by touching it in the course of our detailed inspection of everything within a box or case or enclosure the size of a large radio cabinet, everywhere in every room of every house, every office building, and every factory of every city, and every town of our country.

Conceive the police state that must result from this hard fact in a world from which war has not yet been banished! General Groves, in his testimony before the Senate committee on atomic energy, was asked about the feasibility of international inspection and control. The general was upset at the invasion of corporate and individual privacy that would be involved in making sure that no one was manufacturing atomic explosives, in certifying that the great plants of Oak Ridge and Hanford did not have their illegal counterparts in a country determined to violate the peace. Apparently he did not consider the alternative. Apparently he had not thought of the necessity that, in the absence of adequate international inspection and control, would drive an agent of the FBI to inspect every maiden's hope chest, every matron's china closet, every businessman's file case, every factory's tool cabinet, everywhere in the United States at least once in sixty days. Here is an invasion of privacy to worry about!

Yet such Herculean measures of internal inspection would not be enough. A bomb can be brought in from outside the country in either of two ways. The atomic explosive, which now can be made only in a large, expensive, and easily-identified installation, could be smuggled in little by little by agents, and the rest of the bomb could be built here with the resources of a reasonably modest shop. After all, atomic explosives are respectable-looking metals out of which plated cigar lighters, keys, watch cases, or shoe nails can be fabricated. They cannot be told from other metals except by a detailed study of their density and their X-ray absorption. Here the police state will be needed again. In the insecurity of a world of national atomic armaments, every bit of metal carried by every incoming foreign traveler will have to be inspected in a laborious and sophisticated way.

But we are not yet safe. The other way of bringing in an atomic bomb from abroad is to secrete a complete atomic bomb within some ostensibly innocent item of cargo consigned to the United States. In the hold of a ship floating idly at the Brooklyn docks, awaiting the X-ray inspection of cargo that a fearful nation has imposed, such a bomb could kill its hundred thousand and wreck the harbor. Carried in a plane making its final approach to the landing strip of a great transoceanic air base, it could destroy the base, the travelers, and a few square miles of the neighborhood. The speed and convenience of air travel is nullified by the hard necessity of slow and careful examination of any box big enough to provide for the overseas shipment of a typewriter. A single box of this sort could contain an instrument of enough power to wreck the Panama Canal.

The improved effectiveness of sabotage in the atomic age, the newly increased concentration of destructive energy, make possible anonymous war. The identity of a bombmaker and the names of the men who have planted the bomb will vanish in microseconds in an awesome ball of atomic fire. This opens the horrid possibility of deliberate provocation. In a suspicious world of full-scale atomic armaments, a third nation might, by the planting of bombs, precipitate a war between two others which momentarily fear one another. The treachery that is patriotism in a war is here brought to its utmost depths; yet there is nothing visionary in this thought. It is a grim outcome of what we can do.

It is not likely that a war will be decided by the destruction wrought by atomic bombs secreted in the ways considered here; it is not our purpose to imply that a war can be won by sabotage or betrayal. It need not be; the rockets are too good for that. And the organization of a sabotage network, capable of planting more than a handful of such hidden weapons under the risk of precipitating hostilities if a single plan miscarries, is surely grave. But here is one more uncertainty. If your country is engaged in a race for atomic arms, will you look equably at the great sea of roofs around your house in the city? Any one of them may conceal the bomb. The beginning of a new war will surely involve not only the launching of the missiles, but the explosion of the mines that have secretly been set near key targets to provide the pinpoint accuracy that long-range weapons may possibly lack. Government buildings will fall, the great communications facilities will be destroyed, ports of rail and air and sea traffic will be disabled, the crucial industrial installations will be attacked. All this will happen whether the airborne bombs have pinpoint accuracy or not. We may never know who did it, who planted or smuggled or shipped the bombs. And the measures we shall have to take in an effort to protect ourselves against this assortment of horrid possibilities will reduce international mobility in travel and trade to that which prevailed in the age of sail and will waste our men in guarding, snooping, and investigation.

Our President is reported to be irked with the necessity of having his person closely guarded, day and night. What will some future President feel about the necessity of never approaching nearer than a couple of miles to his fellow man except after the most painstaking scrutiny of the neighborhood and the individuals it contains?

All these possibilities have been predicated on bombs that we know how to build—on bombs our nation is building today against an uncertain future. Like the considerations of other chapters, the considerations here made point only to one fact. We cannot seek national security in armament in a world possessed of atomic arms. Our achievement can only stimulate the ambition and the suspicion of other nations that may be as reluctant as are we to go to war. If one nation arms, all must; and if all nations arm in the terms of the atomic world, each is so overwhelmingly able to destroy the other that a war can almost be regarded as a sanitary measure. Its outbreak becomes inevitable.

The conclusion seems straightforward. We cannot allow our world, beset with many real problems, with much uncertainty and distrust, to drift into a state so fantastic that it beggars words, and so real that you have the photographs of it in the newspapers. An atomic-arms race must be prevented by international control of atomic energy. The saboteur cannot be found, but the factory that makes his bomb need never exist.

FREDERICK SEITZ, JR. *is head of the department of physics at Carnegie Institute of Technology. In the fall of 1943 he joined the Metallurgical Laboratory at the University of Chicago to work on problems connected with the Hanford plutonium plant.*

HANS A. BETHE *lost his position as lecturer in physics at the University of Tubingen in 1933 with the advent of the Hitler regime. He went to Cornell University in 1935, where he is professor of physics. During the war, he was director of theoretical physics work at Los Alamos.*

Chapter 9

How Close is the Danger?

by FREDERICK SEITZ and HANS BETHE

ONE of the views concerning our policy on the atomic bomb that is frequently expressed is contained in the slogan, "Keep the secret!" This immediately raises the all important pair of questions: Is there a secret? If so, can we really keep it in the same sense that we could keep secret our landing position on the Normandy beachhead?

The first part of this query can be answered immediately. At the present moment the British and ourselves possess knowledge of certain basic scientific facts and production techniques that are not generally known throughout the world. These facts and techniques relate to the design and construction of machines for producing pure light uranium (U-235) and plutonium and of bombs made from these materials.

Granting this reply to the first question, we recognize at once that the reply to the second part has tremendous significance in determining our foreign policy. If, for instance, it is beyond the range of possibility to keep to ourselves the facts centering about the technology of the bomb for more than a finite time, such as four or five years, we must pay more attention to our foreign policy than to any other factor in our national program. Otherwise we may find ourselves alienated in a hostile world, a world in which the proximity of sudden death on a large scale is greater than it ever was in the primordial jungle that cradled the human tribes. But before we can answer the second question—namely, how long it will take for another nation to obtain the knowledge necessary to make atomic bombs—we must analyze the history of our own development.

The process of fission, which makes the present atomic bomb possible, was discovered in Germany in the winter of 1938–39 and first became known in this country in January of 1939. This is the starting date of our activity in the field of bomb development that was brought to its practical culmination on July 16, 1945, when the first bomb test succeeded. The six and a half years from fundamental discovery to final application can be broken in more or less clean-cut fashion into three separate periods.

Period 1. The first period extended from January, 1939, to January, 1942, that is, to about the time when Pearl Harbor was bombed. This might be called the period of groping. During it many problems had to be solved: first, the purely scientific problems of devising experimental techniques for investigating the feasibility of a chain reaction and of carrying out the requisite experimental work; second, the problems centering about the procurement of adequate funds and personnel; and third, the problem of maintaining national interest in the issue when the facts were still very hazy and speculative. This was the period when it was most necessary to have men of genius and determination behind the work. It is interesting to note that this was also the period in which the work was carried on by relatively small groups of men at several universities, especially Columbia, Princeton, and California, led by a few of the most brilliant of our scientists. The period ended when it became theoretically certain that the chain reaction would work.

Period 2. This period extended from January, 1942, to about January, 1944. Activities expanded and shifted from pure research to the construction of the first chain-reacting unit and to the design of plants to produce fissionable material on a large scale. Contracts were made with industrial companies to build or operate large-scale plants. Pilot plants were actually built and produced moderate amounts—a few grams—of active materials. These plants were also used to gain further knowledge for the design of the large-scale production plants. Three different kinds of plants were developed, one for the production of plutonium and two for the separation of the isotopes of uranium. The triple development was considered necessary because it was not known which of the methods would succeed, and succeed in the shortest time possible. It caused much additional industrial and scientific effort as well as added expense. In this second period, the Los Alamos bomb laboratory was established, and the design of the bomb was begun.

Period 3. The last period extended from January, 1944, to the summer of 1945. During this time the large plants for manufacturing bomb materials were completed and set in operation. The development passed from the research and pilot-plant stage to large-scale manufacture. The active material produced was used for experiments to determine the size and other features of the bomb, and the bomb design was completed. Bomb research was handicapped by the fact that again several lines of attack had to be followed, just as in the development of production processes. Finally, in July, 1945, the test was made to show the general feasibility of the bomb as well as the soundness of the actual design.

We now reach a point at which it is possible to formulate the problem concisely: How long would it take for foreign countries, other than those involved in the British Commonwealth, to go through each of the three stages described above? The Axis Powers may be left out of consideration. The most important countries are undoubtedly Russia and France, but China or Argentina (or a combination of South American nations) may well be considered too. Also the possibility exists that a highly developed small nation like Sweden or Switzerland might make common cause with a great power of lesser industrial development.

There is no doubt that many of these countries have as much incentive to learn the facts about the atomic bomb as we ever had. Whether a country is justified in any apprehensions it may have is not the problem; in a world of strong national sovereignties, war preparations are made for potential and often remote emergencies. There can be no doubt that in the absence of international control of the atomic bomb, the Russians will try to develop the bomb in the shortest possible time and will devote a large share of their resources to this end. France has already made public the fact that she is starting an atomic-bomb project, with the Sahara as the prospective testing ground and an initial appropriation far greater than that made in this country during the entire first period (1939–1941).

Granting incentive, we may next ask about availability of the scientific talent that will be needed. The United States and Great Britain undoubtedly contain a lion's share of the outstanding scientific talent of the world at the present time. Moreover, this talent has had in the past six years very good conditions under which to work. These two facts explain why we succeeded in developing the bomb in six and one-half years. It would be difficult to argue that any other nation or combination of nations could have done the job faster *if it had started from the same point as we did in 1939.* On the other hand it would be equally difficult to argue that no other country could have accomplished what we have in any period of time. In the first place, Russia and France both have men of outstanding ability. In the second place, it should be recognized that during Periods 1 and 2, in which the major portion of our advance was made, the principal work was in the hands of a small number of people; that is, a large number of good men is not an essential factor. It is almost certain that the reason foreign nations did not proceed very far during the period between 1939 and 1945 (if in fact this is the case!) is because they could not or did not devote full attention to the matter. Russia was fighting for her life with only a fraction of the equipment she needed; France was occupied; Germany, which probably came closest to success, considered the war won in 1941 and 1942 and therefore did not pursue this long-range development with vigor.

What about the availability of materials? The initial work in this country was done on a modest scale with facilities provided by universities. Facilities of this type can be found in all the nations under discussion and many others. Pilot-plant work, corresponding to the developments of Period 2, requires larger quantities of materials, particularly uranium. This mineral is found in appreciable quantity in St. Joachimethal in Czechoslovakia, and also in Russia, Sweden, and Norway, not to speak of the large deposits in the Belgian Congo. We may safely conclude that any nation engaging in the effort will have little difficulty in procuring enough for pilot-plant work.

Considering the ubiquitous nature of uranium it is hard to believe that a nation with the surface area of Russia, for example, could find difficulty in locating deposits of sufficient size to go into large-scale production ultimately. This item might be an obstacle to quantity production of bombs in some smaller countries; however, it would not prevent such nations from engaging in the type of pilot-plant research that is designed to lead to successful plants and bombs—and perhaps trading the results for a price. Moreover, if uranium becomes a material of prime importance to the life of a nation, it will become also more precious than gold, and mining of very low-grade ores will become profitable from the national point of view. Gold is extracted commercially from gravel containing as little as 0.3 part of gold per million; the average uranium content of the earth's crust is believed to be about twenty times as large as this, or 6 parts per million.

With respect to industrial capacity, many of the countries mentioned are far advanced. True, the scale of their production is not so large as ours, but 2 billion dollars is by no means an excessive sum for any of them in comparison with their national income in, let us say, five years. *Moreover, as we shall discuss below, it will probably be far less expensive to repeat our development a second time.*

Some data collected by Dr. Lawrence Klein of the University of Chicago show that in Sweden—to take one of the smallest of possible bomb-makers—the average annual gross output of plant and equipment (that is, nonconsumption goods) in the period 1925-1930 was 350 million dollars. Much of this was used for replacement of old plant and equipment, but when the issue is one of preparation for war, this factor can be readily modified. During our own war-production years, we managed to supply sufficient armaments to the United Nations armies by not making the customary replacements of our plant and equipment. Sweden, in the act of producing bombs, could also consume capital in order to reach her objective.

An all-out effort on the part of Sweden might call for an average expenditure of 200 million dollars per year for five years. In terms of her 1925–1930 output, this would absorb 57 per cent of her capacity to produce plant and equipment and only 10 per cent of her gross capacity to produce goods and services of all types. These percentages are very small as compared with those required by the war efforts of the United States, the Soviet Union, Great Britain, Germany, etc. Such a program would be a rather simple order for Sweden if she really wanted atomic bombs.

Many Americans believe that the Russians, although capable of quantity production, are backward in the quality of their industries. The writers should like to point out, as just one item bearing on this point, that the Russians carried out an extensive tank program during the war and produced in quantity a tank that was as good as the best German production and, in the writers' opinion, far better than our own Sherman. This program must have involved an effort comparable to that which we put into the atomic bomb, from the standpoint of both technology and production.

In coming to an estimate of the time required for a foreign nation to produce the atomic bomb, we must compare not only its resources with ours but also its starting point with our starting point. Any nation which begins working on the development of the atomic bomb at the present time starts with far more knowledge than we possessed in 1939. The two major sources of information are, first, knowledge that the bomb works and is sufficiently small to be easily transported by air; and second, the Smyth report which contains some rather specific data.

Consider first the advantages derived from knowledge that the bomb works. Such knowledge means that much of the groping and speculation that was necessary during Period 1 of the three periods discussed above is unnecessary. Thus, the incentive for working very hard and on a large scale from the start is provided immediately. The greatest effort of Period 1 in our development was devoted to obtaining scientific aid and financial backing, and all this time can now be saved. Moreover, it is no longer necessary to depend upon the vision and judgment of the men of rare genius. Quite average scientists can appreciate the factors involved. Further, it becomes possible to reduce the total time by starting all three phases of the program at once. It is no longer necessary to await the results of the first period of development before deciding how much effort to risk on the second and third periods.

Consider next the Smyth report, which provides detailed qualitative information on the general direction in which work can profitably be pushed. With respect, for example, to the production of plutonium, the Smyth report states that one can expect to operate a reactor pile with the use of natural uranium and a graphite moderator, even though water, which absorbs neutrons and thereby cuts efficiency, is allowed in the system as a coolant. Moreover, it follows from the report that plutonium produced during the reaction can be separated chemically and in sufficient purity for use in a bomb. The report does not furnish precise information on the dimensions of the pipes and other conduits used in the installation, nor does it describe in detail the methods used for chemical separation. However, men of a far lower order of genius than those who planned the original work could undoubtedly fill in the missing pages as long as they are bolstered with the positive knowledge that the entire program is feasible.

At a dinner table conversation during a recent scientific meeting, a competent physicist who had not been associated with any of the developmental work on the bomb was heard to relate to another physicist his inferences concerning the production of plutonium and the general dimensions of the atomic bomb, all of which he had drawn from reading the Smyth report. The agreement with unpublished facts was remarkable. And it is safe to say that there are at least twenty-five people in each of the foreign countries under discussion who could infer as much as he did from the Smyth report.

Equally important is the information that any one of three different processes will lead to success—the production of plutonium by the chain reaction, the separation of uranium-235 by the electromagnetic method and separation by the diffusion method. Any country starting now with this knowledge can determine with relative ease which of these processes is the cheapest and most adapted to its own industrial facilities. That means a very great saving in money, probably reducing necessary expenditures well below 1 billion dollars. It means a substantial reduction in industrial and scientific effort because all work can be concentrated on one line.

What reduction in time results from all the knowledge now available? The greatest reduction will, of course, occur in the first period of development. This period required three years for our own groups, who worked for a large part of this time without much financial support and without the knowledge of ultimate success. Having this support and the information of the Smyth report, it is difficult to imagine that men of the quality of Auger and Joliot in France and Kapitza, Landau, and Frenkel in Russia would require as long as we to cover the same ground. Two years could easily be sufficient for this period.

Regarding the second phase of the work, we can say with safety that there is now no risk in beginning the plans for pilot-plant operation at once. Detailed data for the production of such units may not be available at the moment. However—if, for instance, it is decided to produce plutonium rather than to separate U-235—it is known that uranium and graphite will have to be used in quantity. As a result, work on the preparation of these materials can begin at once. Here again the Smyth report is very helpful because it indicates that a certain rather simple process for the manufacture of uranium metal has been successfully used—a fact which took long to establish in our own development. Similar preparatory work can be done if one of the separation methods is selected as the most suitable process. In either case, a site can be chosen immediately for the ultimate location of the pilot units, and all necessary preparations for servicing the units can be started. Thus, perhaps a year after research in Period 1 has indicated the dimensions to be used in the pilot units, these units can be functioning.

We come next to the question of large-scale manufacture. All the reasoning that has been applied to pilot-plant production can be applied here too. The proper sites for processing and purification of materials such as uranium and graphite can be carried along with the corresponding work for the pilot units. Some delay might be caused by the development of a chemical process to separate plutonium from uranium, because such a process can probably be found only after the pilot plants have produced sufficient material to work with. Moreover, at this stage the high development of industry counts most, and other nations may require more time than we did because their industry is either in quality or in quantity behind ours. Even so, we are probably putting the figure high if we allow two years for this period, which is about twice as long as the time we required. Adding this to the three years estimated for the completion of periods 1 and 2, we conclude that manufacture of plutonium or uranium-235 (or both) can be under way in five years at the outside. It is clearly recognized, of course, that final manufacturing can be carried out only by a nation that has suitable sources of uranium.

Finally, we come to the all-important issue of the design and construction of the bomb. The design can start very early in the program, probably relatively earlier than in our own development. Basic information obtained during the first two periods of development, through the pilot-plant stage, will provide the necessary knowledge about the bomb dimensions and methods that can be used to detonate it. This information should be available in about the fourth year, according to our estimate, so that the theory of the bomb will be clearly understood by the time the manufacturing units are beginning to yield the bomb material. With much of the bomb design done in advance, it is unlikely that there will be any major delay between manufacture of material and production of the finished product. A year is certainly an outside limit. Altogether we have, therefore, a total elapsed time of six years before bombs are available—slightly less than the time needed by us, in spite of the fact that we have added a year to take into account the supposedly lesser industrial development of other countries.

Thus we find that other nations will probably be able to duplicate our development in about the same time that we required. The Smyth report, valuable mainly because it states that certain processes that would naturally occur to other scientists were actually successful, will be a considerable help in their program. The most important fact, however, is that our entire program was successful. Even if this bomb had not been demonstrated, it would soon have become known that three major factories were engaged in our program, of very different appearance and with very different machinery, and that all continued to operate, showing that there must be three different successful processes. Much of the most relevant information in the Smyth report could thus have been deduced from other evidence. The main secret was revealed, and the main incentive supplied, when the first atomic bomb was dropped on Hiroshima.

Many factors can enter in to reduce the required time estimated here. For one thing, we have adopted all along the somewhat provincial viewpoint that the nation engaging in the work will be less effective than we have been, and this viewpoint may be entirely unjustified. Also, it should be kept in mind that work in one or another nation may already be much farther along than external facts would indicate. Finally, it must never be forgotten that men of genius in other countries may devise methods which are much superior to our own and which would greatly reduce the time involved; our previous estimates have been based on the assumption that a foreign nation would simply copy our own pattern of attack.

To summarize, then, we are led by quite straightforward reasoning to the conclusion that any one of several determined foreign nations could duplicate our work in a period of about five years. The skeptical or nationalistic individual might at this point decide that such reasoning should have little effect upon our foreign policy, because it is possible that in five years we shall be so far ahead of our present position that it will not matter whether or not a foreign nation has our present knowledge.

There are two very grave objections to this viewpoint. In the first place, it is entirely possible that a foreign nation will actually be ahead of us in five years. In the second place, even if we have more powerful bombs than they, our preferred position will be greatly weakened. For it is an unfortunate fact that present bombs are of sufficient strength, if used effectively and in sufficient quantity, to paralyze our highly centralized industrial structure in the space of a single day. Any store of more powerful bombs in our arsenals would be of little value unless we could use them to prevent attack, and this seems to be a very remote possibility. The existence of such bombs might have an inhibiting effect in the sense that the enemy would fear reprisals. However, if history provides any lesson, it is that fear of reprisal has never prevented a war in which the chances for quick victory are as great as they would be if an adversary decided to strike rapidly and in full strength with atomic bombs.

IRVING LANGMUIR, *one of America's most distinguished industrial scientists, was awarded the Nobel Prize in chemistry in 1932. He is associate director of the General Electric Research Laboratory. In 1945 he was one of a small group of U. S. scientists who attended the anniversary meeting of the U.S.S.R. Academy of Sciences.*

Chapter 10

An Atomic Arms Race and Its Alternatives

by IRVING LANGMUIR

WE now have atomic bombs and are accumulating materials that could be used for making them. This program is going ahead with a yearly expenditure of roughly 500 million dollars. It has been announced that Great Britain is planning to produce atomic bombs. On November 6, Molotov said, "We shall have atomic energy, too, and many other things."

An atomic armament race has thus started that brings insecurity to all nations. Yet every one of the United Nations desires future security more than almost anything else. International control of atomic energy and of materials used for atomic weapons is thus of the utmost urgency. If a method of control is not worked out, the only alternative seems to be the development of an atomic armament race which will undoubtedly end, as all previous armament races have ended, in war.

Stages in an Atomic Armament Race. I shall attempt to analyze the successive stages in such an armament race. In the first stage the United States alone will have atomic bombs and will accumulate a stockpile. Other nations will be preparing to make them. During this time we are in a secure position. In the second stage one or more other nations will have begun to produce atomic bombs while the United States stockpile may become so great that we will have enough bombs to destroy practically all the cities of an enemy country. During this period we are still relatively secure. During the third stage many nations will have enough bombs to destroy practically all the cities of any enemy. During this stage no nation is secure. Since an attack by any nation would almost certainly be followed by retaliation, any lasting advantage of a surprise attack largely disappears.

If an atomic armament race continues long enough, it is probable that discoveries will be made by which the production cost of the bombs may be greatly decreased, or new types of bombs may be devised thousands of times more powerful. It has been estimated that about 10,000 bombs of the present type might destroy nearly all the cities of the United States. The area covered, however, would be about 100,000 square miles, which is roughly 3 per cent of the area of the United States.

During the fourth stage of the armament race, atomic bombs or radioactive poisons distributed over the country might destroy practically the whole area of the country, so that no effective retaliation could be offered. The victor in such a war would then have to dominate the whole world so effectively that he could not be endangered by other atomic bombs. The fourth stage in the atomic armament race, if it is allowed to proceed that far, will bring intolerable insecurity to most nations, so that the nation which feels that it is best prepared is almost forced to start a war to avoid danger of complete destruction.

The rate at which nations can progress through the four stages of the armament race depends not only on the difficulties inherent in atomic-bomb production but also, in very large part, on the motivation: Just how much effort do the various nations consider they can afford to make to attain their objectives?

The incentives that would lead to the effort are of two kinds: first, questions of prestige; second, the intensity of the feeling of insecurity. Such insecurity will probably ebb and flow according to the international situation. The fact that we are now the only nation possessing atomic bombs means that during the early stages in the armament race other nations will act largely according to their understanding and interpretation of

American intentions. It is thus of particular significance that in the Truman-Atlee-King declaration of November 15, 1945, it was recognized that the United States, the United Kingdom, and Canada should "take an initiative in the matter."

Possible Atomic Developments in the USSR. Undoubtedly, Great Britain and Canada are the nations, outside the United States, that could first build atomic bombs. Churchill has already said, "It is agreed that Britain should make atomic bombs with the least possible delay and keep them in suitable safe storage."

But Russia, with her population of over 195 million in an area of about 9 million square miles, also has enormous resources in men and materials. During the years 1934–1940, Russia, instead of following a policy of appeasement like that of other nations, engaged in a vast program of military preparation not for purposes of aggression but for defense against German aggression. They did this even though it meant holding the standard of living far below what would have otherwise been possible. The military experts in Germany and America greatly underestimated the Russian military preparation, and they were taken by surprise at the power that Russia showed in driving back the German armies from Stalingrad to Berlin. The Russians built remarkably good planes, for many years holding the world's record for long-distance flight. The efficiency of Russian tanks compared with those of Germany and the United States has already been pointed out in the preceding chapter.

The cost of an atomic-bomb project like our own would be small indeed compared to the expenditures that Russia made in preparations for the recent war. Since atomic bombs are roughly ten times cheaper than other weapons in terms of equal effectiveness, the over-all cost of even a large atomic-bomb program might be much less than Russia would normally expect to devote to an army and navy of the conventional types.

The Russians give the impression of being a strong, rough, pioneering people who are proud of their accomplishments during the recent war. Questions of prestige would, therefore, probably play a strong part in stimulating them to learn to control atomic energy. If the international situation develops in such a way that they feel increasing insecurity, I believe that the Russians might launch a program for the development of atomic bombs on a far greater scale than that likely to be undertaken by any other country. Russia can mobilize her resources for such a program just as she did in her preparations for the war with Germany, devoting 10 or even 20 per cent of her capacity to a five- or ten-year plan. Before the war the United States spent only about 0.04 per cent of its national income for research in pure science and 0.25 per cent for industrial research. During the war total research expenditures, including the atomic bomb project, rose to perhaps 1.5 per cent.

In such an extensive project the efficiency of the Russians might at first be low, but this would increase rapidly and steadily as they progressed with their plan, just as it has in all of their major undertakings. They are quite used to big projects. When I was in Russia recently, I was told of a pilot plant, costing nearly 100 million dollars and nearing completion, for the continuous operation of a large blast furnace using oxygen instead of air. Experimental runs that were made before this plant was designed proved that a blast furnace of a given size gave about five times the output when oxygen was substituted for air. A 2-billion-dollar project was under consideration for converting the whole steel industry of Russia, which would result in a large saving in the cost of steel and iron.

If an armament race continues, I believe the Russians may reach stage two (that is, they will have begun to produce bombs) within about three years. Thereafter, however, there is a definite possibility that Russia may accumulate atomic bombs far faster than we do, so that they may get to stages three or four even before we do. The advantages they have in such a race are

1. They have a large population; it can be regimented and is willing to sacrifice living standards for a long-range defense program.
2. They have a remarkable system of incentives, which is rapidly increasing the efficiency of their industrial production.
3. They have no unemployment.
4. They have no strikes.
5. They have a deep appreciation of pure and applied science and have placed a high priority on it.
6. They have already planned a far more extensive program in science than is contemplated by any other nation.

Russian Scientists. The rapidity with which a large-scale atomic-bomb project could develop in Russia, given incentives of the kind mentioned above and failing control mechanisms, would ultimately depend upon her ability to train scientists. It has recently been stated that there are now in Russia 790 universities and that the number of students has been increasing steadily in spite of the war. They believe that their educational methods have improved very greatly. I was told, for example, that they discovered during the war that they could train skilled workers for industry in a far shorter time than was previously thought possible.

Even so, many people have believed that Russia has not a sufficient number of scientists nor the educational facilities for the training of scientists and has not sufficient skilled labor to build atomic bombs within a reasonable time. General Groves, for example, in his testimony before the Atomic Energy Committee of the Senate, thought it might take Russia as long as twenty to sixty years to build atomic bombs.

I had an opportunity to become familiar with some branches of scientific development in Russia by attending the meetings in Moscow and Leningrad in June, 1945, held in commemoration of the 220th anniversary of the founding of the Academy of Sciences of the U.S.S.R.

The plenary meetings of the Academy, which were held in large opera houses with about 3,000 people present, were devoted to general papers on the history of science in Russia and in other countries and on selected subjects of wide general interest. More than 100 foreign scientists had been invited to attend these meetings. Most of our time during the eighteen days in Moscow and Leningrad was spent in conferring with scientists in some of the seventy-eight institutes of the Academy. I visited several institutes, particularly in the fields of chemistry and physics. I found that the Russian scientists talked freely about their work and showed me through their laboratories. I was much impressed by the friendliness of all of these men and by their wholehearted devotion to science. They were all clearly working on problems that had been planned by scientists without undue political control. In fact, they had been able during the war to carry on scientific work of a kind that would have been impossible in the United States. A great deal of the work was of long-range character, often planned to lay sound foundations for postwar industrial developments. They had been able to defer men from active military service for such work.

Among the men whom I met, there was clearly a desire for a long period of peace and security. Their plans indicated that they hoped and believed that this would be possible. During the years from 1934–1940 there had been a great feeling of insecurity, for they said they all felt the danger of German aggression. In June, 1945, they showed great relief at having reached the end of this period of insecurity: The war against the axis powers had been won. They were planning to repair the damage in the devastated areas but expected at the same time to lay the foundations for a future standard of living as high as or higher than that in the United States. Science, pure and applied, was to play a dominant role in this program. The building of the Academy of Sciences had recently been renovated and improved, but I was shown plans for a new building at least five or ten times larger than the present one.

The social position of scientists in Russia, as well as the provision of summer homes, automobiles, etc., are held out as incentives for men to become leaders. At the meeting of the Academy about 1,400 honors were given; for example, 13 scientists received the Hero of Socialist Labor, the highest honor; 196 received the Order of Lenin, which only a few weeks ago was given to Molotov. An article entitled "Science Serves the People," which appeared in the *Moscow News* at this time, contained the following:

> . . . Never before has the scientist been accorded such attention by the state and such esteem by society as in the Soviet Union . . .
>
> . . . The state provides the maximum amenities for life and facilities for work to the scientist and assures a comfortable life to his family after his death . . .

It is extraordinary that this meeting of the Academy was held only about a month after the end of the war in Europe. The lavishness of the entertainment—for example, a banquet for 1,100 people in the Kremlin with Stalin in attendance and Molotov as toastmaster—showed the importance the Russians attach to science. In all the speeches great emphasis was laid on the international character of science. It was stated that scientists the world over had always cooperated with one another, national antagonisms playing no role. Hope was expressed that in other fields nations might learn to cooperate in a similar way.

The use of the atomic bombs in August against Japan must have come as a great shock. Most of the Russians probably felt that the security that they thought they had reached was suddenly ended, and they were brought to a state of insecurity like that of the years 1934–1940. I believe that the difficulty in reaching international agreements with Russia before the Moscow conference was caused by a natural reaction arising from their disappointment regarding future security.

We can better understand Russian doubts about our policy of holding atomic bombs as a "sacred trust" by asking ourselves: What would American public opinion now be if we had had no atomic-energy development, but if near the end of the war, atomic bombs had been dropped on Berlin by the Russians without adequate consultation? Would our insecurity be entirely relieved if the Russian government, a few months later, had announced that it held an increasing stockpile of atomic bombs as a sacred trust?

Basis of Agreement with Russia. The declaration of November 15 holds out the hope that through the United Nations Organization there may be established "an atmosphere of reciprocal confidence in which political agreements and cooperation will flourish".

A basis for cooperation and for ultimate world control of the atomic bomb may be sought by considering, first, matters of common agreement: No nation desires world conquest, each desires security, freedom from unemployment, better conditions for labor, and, in general, a higher standard of living for its people. Many other points of agreement can be found.

There are serious difficulties in world control. We must take into account that Russia and the United States do not understand each other very well. We don't like their form of government and they don't like ours. They don't like our strikes and unemployment, and we don't like their control of the press and public opinion. They find many statements in our newspapers that they know are false, but apparently we wish to believe these statements. The Russian newspapers attribute everything they don't like about America to the control by capitalists, plutocrats, or the bourgeoisie. They say that they have the only true democracy, but we say they have no democracy at all. Under such conditions it is hard to allay fears or distrust.

Nations differ fundamentally in their forms of government. They can agree, however, to recognize the right of each to its own form of government within its own borders. Disagreements have arisen of late regarding the governments established in conquered and liberated countries.

General policies for the United States have been proposed in the Four Freedoms and in the Atlantic Charter. The first of the Four Freedoms given by President Roosevelt in January, 1941, is "freedom of speech and expression . . . everywhere in the world." In the Atlantic Charter proposed by President Roosevelt and Prime Minister Churchill there are eight articles. The second article states that our nations "desire to see no territorial changes that do not accord with the freely expressed wishes of the people concerned." The third declares that they "respect the right of all people to choose the form of government under which they will live," and they wish "to see sovereign rights and self-government restored to those who have been forcibly deprived of them."

There is a general American belief, which makes it hard for other nations to cooperate with us, that world problems can be solved by slogans or idealized principles even when these are really not applicable to concrete situations. The troubles caused by applying some of our ideals to other nations can be understood by considering the differences in meanings attached to democracy and the freedom of the press in the United States and in Russia.

In referring to democracy the Russians are apt to use a standard phraseology. A good example is given by the following quotation from an article entitled "Soviet Democracy," by Professor V. Baushko, published in the *Soviet News,* November 3, 1945:

> There can be no consistent democracy in bourgeois countries—not even in those of them whose constitutions proclaim these rights. Wherever society is divided between the exploited and the exploiters, there can be no equality. . . . There can be no freedom of speech, press or association for the toilers if print shops and paper and even meeting halls belong to the bourgeoisie. . . .
>
> In the USSR, on the other hand, all governmental

> power belongs to the people, exploitation of man by man has been done away with, there are no classes of exploiters, democratic rights and liberties are guaranteed by the Socialist economy . . . racial and class hatred has been eradicated and supplanted by friendship in relations of the peoples, while science and culture have been placed at the service of the people.

We believe that in the Balkan states and in Japan the peoples should have democratic forms of government, and Russia seems to agree to this. However, in Japan we pattern the proposed democracy after our own, but we do not agree that Russia should pattern the governments of the conquered and liberated Balkan States in accordance with her conception of democracy. If we are to get along with other nations that have different ideals, we cannot insist that our concepts of freedom and democracy shall prevail everywhere. Such matters involve compromise and wise statesmanship. They are not to be settled by means of slogans.

The freedom of the press is another point in which we differ from Russia. The following is a quotation from Molotov's report of November 6, 1945, on the occasion of the twenty-eighth anniversary of the October Revolution:

> The strength of the Soviet system lies in its closeness to the people. As distinct from parliamentary democracy, Soviet democracy is truly popular in character. Therefore, the Soviet state, as a state of a new type, has tasks which are not inherent in states of the old type. Thus, the duties of the Soviet state include the political education of the Soviet people in the spirit of safeguarding the interests of world peace, in the spirit of establishing friendship and cooperation among the peoples, which . . . calls for the exposure of all attempts to prepare a new aggression and the regeneration of fascism. . . . Under the Soviet Constitution it is a crime to preach hatred among nations, antisemitism, etc., just as praising of crime, robbery, and violence against man is forbidden in our press. Such "restrictions" are as natural under Soviet democracy as things quite opposite are unfortunately natural for some other states. In some countries freedom of speech and press are still interpreted in such a way that mercenary servants of fascism do not even have to don masks in order to carry on unbridled propaganda for aggression and fascism . . . .

It is clear that the Russians have a radically different concept of the freedom of the press from that which is current in America. We certainly cannot demand, in accord with the first of the Four Freedoms, that our kind of freedom of the press shall exist "everywhere in the world." The Russians would see no reason why their ideas on this subject should not apply to the Balkan states if ours are to prevail in Japan.

Americans often have no confidence in the Russian press because it is government controlled, but we fail to recognize that a great deal of distorted news is introduced into our papers even by propaganda agencies with malicious intent. A good example was the statement in our papers, reiterated by Secretary Stettinius at the San Francisco Conference, that sixteen Poles representing the Polish government in London had been invited to Russia for a conference and had then been arrested and were being held for trial. This statement was later denied by Stettinius. Ambassador Harriman in Moscow told us that the sixteen Poles had not been invited to Moscow for a conference but were arrested in Poland for distributing arms for use against the Russians. I attended some of the sessions of the trial of the sixteen Poles. I believe that they had a fair trial. Many of them were acquitted. The maximum penalty was ten years' imprisonment. The defendants were proud of their actions against Russia, and one of them said that he was willing to fight against Russia if necessary to help Poland acquire an outlet to the Black Sea. I have never found anyone in America who had heard the denial of the original false story. It was probably published inconspicuously. Such unfairness produces a bad effect on our relationships with Russia.

It is highly desirable that there should be frank discussions between the Russian government and the American government regarding the troublesome effects caused by differences between our concepts of democracy and freedom of the press. One cause for the lack of mutual understanding between the United States and Russia is the fact that so few people of each nation travel in the other. It was pointed out recently by Edgar Snow that there are only 260 Americans in Russia and about 2,000 people carrying Russian passports in the United States. The facilitation of travel between the two countries would help greatly. It is encouraging to find that Molotov in his address of November 6 recognized the desirability of this intercourse. He said, "Acquaintance with the life of other nations would certainly be a benefit to our people and would broaden their outlook."

Removal of restrictions on the circulation of American newspapers and periodicals in Russia and Russian articles in the United States will make for much better relations. We must remember that in our country anyone who talks too much about Russia or approves of Russian practices is liable to be "investigated" by the House Committee on Un-American Activities. A very hopeful sign was Russia's recent invitation to newspaper correspondents to visit all parts of Russian-occupied territory.

One of the recommendations made in the November 16 declaration was that we should exchange scientists and scientific information among nations. Russia started such action in June, 1945, by inviting about 120 foreign scientists to Russia and giving them full information regarding scientific work.

Russia publishes a majority of its important scientific papers in English as well as in Russian. They have started to teach English to all school children in Russia and even in Siberia. Russians are inveterate readers and, if they can read English books and journals, a big step toward better understanding will be reached.

Problems of Inspection. An interchange of scientists would also pave the way for effective methods of inspection, which will probably be needed for an effective world control. Dr. Szilard writes about this in detail in a later chapter, but I should like to note some particular aspects of it here.

I believe that if there is a sincere desire for security on the part of all nations, there will be a universal insistence on an effective inspection system. This could involve the inspection of sources of uranium, and, of course, the inspection of factories making materials used for atomic energy.

It was pointed out by Bernard Brodie (*The Atomic Bomb and American Security,* published by the Yale Institute of International Studies, 1945) and again emphasized strongly by Dr. Urey in his testimony before the Senate Committee on Atomic Energy that inspection would be very much facilitated if large-scale plants for making U-235 or plutonium for power purposes did not exist.

The greatest peacetime benefits to be derived from our new knowledge of nuclear reactions will probably come from its indirect effect in speeding up progress in science leading to great discoveries in biology, chemistry, and physics. These benefits can probably be obtained by small-scale production of radioactive substances by one or two piles. Large-scale use of atomic power would involve the production of materials that could be quickly converted to use in the atomic bombs. This would present very serious inspection problems. The commercial use of atomic power as a substitute for coal and oil will for many years be a matter of trivial importance as compared to the dangers that might result in the existence of atomic bombs.

It would, therefore, be desirable to destroy all atomic bombs, all large plants for making them, and all reserves of elements of U-235 and plutonium if these should prove serious stumbling blocks in reaching effective world control and necessary inspection.

It has been proposed that stockpiles of atomic bombs should be turned over to the Security Council of the United Nations Organization. It is hard to conceive that a reserve of atomic bombs would serve any useful purpose. The atomic bomb is not a police weapon.

If a step-by-step process finally brings the envisioned effective control of atomic weapons, world confidence will grow and a mechanism will then exist by which other weapons may be outlawed or controlled. We may some day come to regard the atomic bomb as the discovery that made it possible for mankind to bring an end to all war.

Chapter 11

HAROLD C. UREY *was awarded the Nobel Prize in chemistry in 1934 for his discovery of heavy hydrogen. He became a member of the Uranium Committee in June, 1940, and from that time directed the project work on the separation of U-235 by the centrifugal and gaseous diffusion methods at Columbia University. He is now at the University of Chicago.*

How Does It All Add Up?

By HAROLD C. UREY

DURING the nineteenth century the elements and techniques of mass production were developed. They have been improved and exploited for peaceful purposes during this century and especially in this country, until today nearly everything we use is produced by mass methods. Without these methods the high standard of living in the United States would be impossible, and with them a high standard of living is possible in all countries of the world that have reasonable resources. During the present century many scientific and engineering discoveries have been made that contribute to this high standard of living.

Also, unfortunately, mass-production methods and scientific discoveries have been used for purposes of war. In order to apply these methods and discoveries to a given purpose, we must have an opportunity to practice them. The First World War gave the first opportunity for adaptation of mass production to purposes of war. But it was only an elementary course in the art of mass destruction. The Second World War gave the opportunity for the advanced course, and by the end of it the lessons were thoroughly learned. Today we have the scientific knowledge, the engineering talent and experience, and the industrial know-how to make war on a real mass-production basis. Another war would differ from this past one in the same degree as a modern automobile differs from the Model T Ford, or perhaps a horse and wagon. Another war would be so successful from the point of view of destruction that little of the physical and human bases of our civilization would be left. For our scientific and mass techniques now include the atomic bomb and probably other weapons not yet come to the attention of the general public.

The particular weapons that concern us here include the airplane, the pilotless flying bomb (V-1), the rocket bomb (V-2), and the atomic bomb. It is possible that other methods suitable for delivering the atomic bomb may be partially developed and as yet unknown to the public. Still other methods may be developed in the future. Only the B-29 bomber was used in this war for the delivery of the atomic bomb, but the combination of these two weapons made conditions intolerable to Japan. In the future these weapons, produced on a mass-production basis, will make war intolerable to all peoples of the world. This does not mean that war will not come. It does mean that the war will probably not be of long duration because of the vast destruction that will be quickly and decisively accomplished.

Let us review the facts in regard to the atomic bomb as we see them today and as the preceding chapters of this book have laid them before us.

The atomic bomb, because of its overwhelming increase in effectiveness, which makes all defenses known or forseeable nearly useless and completely ineffective, is not best regarded as just another weapon. In the past many new weapons have been invented, and in many cases they have added greatly to the effectiveness of the attack as compared to the defense. But our present defenses against atomic bombs are about as effective as a Roman army armed with spears, javelins, and shields would be against a modern army equipped with machine guns. Within a few years the atomic bomb carried by modern airplanes has increased the advantage of attack by about as much as was accomplished in a thousand years in the past. These weapons can destroy all defenses that we can devise at the present time.

Dr. Ridenour, in Chapter 7, has reviewed in detail this question of defense against the atomic bomb. Still, many people believe that there will always be a defense for every weapon. There will always be an exception to such glib rules. But is the statement true in any sense that is of interest to us? Is there a defense against bullets? Perhaps, but they killed many men in this past war. Is there a defense against submarines? Yes, definitely. But they destroyed a large fraction of the world's shipping during this past war. Is there a defense against airplanes? Certainly defenses are known, but only the United States among the major combatants of the Second World War escaped serious damage or nearly complete destruction of its cities. Similar answers to similar questions could be made with respect to tanks, naval vessels, and other weapons great and small.

Weapons disappear from war if they are superseded by more effective weapons, but so long as they are used in war they produce actual damage in spite of defenses and in a rather definite proportion to their ability to inflict damage. When more destructive weapons than atomic bombs are developed, atomic bombs will not be used; but so long as they are used, they will continue to destroy many square miles of cities for each bomb exploded. Perhaps a question points the argument. Can any of us imagine a defense so effective that, sometime in the future, a country such as the United States would decide not to manufacture atomic bombs because defenses against them made their manufacture inadvisable for purely military reasons? I think not. If atomic bombs are not made in the future, it will be for other reasons than the effectiveness of defense. No military defense exists, and none can be devised. Atomic bombs are able to destroy the cities of the world and they will do so, if used in another war.

This thesis assumes that atomic bombs can be made in sufficient numbers and at costs sufficiently low to be used effectively in another war. Unfortunately both assumptions are correct. Our mass-production methods, which give us our high standard of living, our automobiles, power plants, chemical products, electrical devices, etc., make possible the production of atomic bombs in large numbers and at low cost. In fact, war will be cheaper in the future so far as the production and use of weapons are concerned and far more expensive from the point of view of destruction accomplished. Even small countries can make these bombs in numbers if they are such utter fools as to engage in the lethal business. They will not, because they know that they would be completely destroyed if bombs were used. It is only powerful industrialized countries that may not realize that before this weapon all countries are small and weak. Before the past war small countries realized that they must get along with their neighbors. The cheapness of the atomic weapon relative to its destructive power makes it necessary that *all* countries get along with each other.

The question is asked: Can other countries besides England and the United States produce these bombs? And the answer is: Of course they can. What weapon ever devised by man remained the sole possession of the country of origin? The production of atomic bombs was a tricky, intricate business, but so is the production of tanks, airplanes, and other major weapons of war. The United States is the greatest industrial power at the present time and could and did make these weapons faster than any other country. But it is nonsense for us to assume that other countries cannot learn all the details of their production and, in fact, improve on the methods. If the people of the United States or England believe otherwise, they engage in the most dangerous of delusions.

How long will it take other countries to devise these weapons? Estimates vary. Most scientific and technical men who helped to produce the bombs guess between five and ten years; a few think less and some think more. Drs. Seitz and Bethe give convincing reasons for their estimate of six years or less in Chapter 9. It is to be hoped that the time will be large rather than small, since that would give more time for a solution of the whole problem.

Some have suggested that the United States would be safe if it were to keep ahead of other countries in the development and production of atomic weapons. It is not certain that this country could succeed in this attempt over any great length of time and certainly not for all time, for others will surely come abreast of us—and perhaps in less time than we think. But let us look more closely at the suggestion. Supposing that the United States remains ahead of other countries in number or effectiveness of bombs, what good does it do us? Do we plan to attack other countries at a favorable moment? Following such an attack, it would be necessary to occupy the countries with our armies in order to prevent the manufacture of bombs in the future. Some 7 per cent of the world's population would have to keep its feet on the necks of the rest of the world's peoples. It does not seem likely that we would choose this role voluntarily, with a full understanding of its responsibility and hardships let alone other considerations.

Later, when other countries secure enough atomic bombs to destroy the cities and other appropriate targets of this country, we would be in a position to destroy their cities, and it would do us no good whatever to have bombs enough to destroy those targets more than once. Extra bombs would be useless once sufficient numbers were available to destroy all large military targets of any possible enemy. If our hypothetical enemy had sufficient bombs to destroy our military targets, in what way could we keep ahead of this enemy? Atomic bombs are different. Enough can be made to destroy completely all possible targets and kill the inhabitants of all major cities of any country. It is then impossible to destroy them twice or to kill people twice. Eventually, therefore, we cannot hope to keep ahead of other countries in an atomic war.

We turn to defensive measures again and specifically to the dispersal of cities and going underground. This appears to be the only effective defense that anyone has proposed as yet, and it is only a palliative, only a way of moderating the effects of an attack. The cost of such dispersal would be high, on the order of the cost of the Second World War to us, for the actual translocation of dwellings, industries, and transportation facilities. Dispersal would also impair the efficiency of our industrial system, since industries are usually placed where they are because of economic advantages, such as natural transportation facilities and availability of power or raw materials. In many manufacturing processes concentration is often of great advantage; for many industries it is necessary.

The psychological problems associated with the dispersal of cities would be great. Many of us, and perhaps most of us, for one reason or another like the places and conditions under which we live, though you and I may not understand why others do so. Dispersal would affect each of us in a very direct manner. Agreement to carry through such a program would never be unanimous, and possibly the decision in regard to it would rest on a mere majority, with a determined minority vigorously opposed to it. Proposals have been made that dispersal could and should be achieved in fifteen years; this seems too short a time, though in the face of the threat under which we all may live, we may decide to carry it through. It would probably require a dictatorship for its execution. And in the end it would not be a definite and decisive defense, for if somewhat larger bombs were secured a determined enemy could destroy our economy and our people just the same. The atomic bomb is a very effective and inexpensive weapon and will probably become more effective and less expensive with time.

Perhaps critical plants could be placed underground, but to what purpose? This would not prevent the destruction of the people above ground, and if military plants and installations cannot protect the citizens of a country, what are they for and why prevent *their* destruction? If the Navy, Army, Air Forces, and atomic bombs cannot protect the citizens of a country and their property, who cares whether or not the military forces can protect themselves?

This book is concerned primarily with the military threat of atomic energy rather than with its peacetime uses, but such possible uses have a bearing on the military applications. Radioactive materials for medical uses can be secured without large plants containing large amounts of fissionable materials, for example, U-235 or plutonium. However, atomic power plants must necessarily contain sufficient of these materials to make bombs. Undetected diversion of these materials for use in bombs might be comparatively easy, since all chemical plants lose some material and losses vary from plant to plant. Records could be falsified comparatively easily, if those operating the plants were determined on such falsification. A much more extensive inspection system would be needed to prevent diversion of material in operating plants than would be necessary to prevent the construction of these plants. The possibility of diversion would not be conducive to confidence on the part of the peoples of the world. And since confidence in the operation of world control of atomic bombs is so vitally important, everything should be done to promote it first. If proper controls could be secured and confidence established, the operation of power plants could *then* be considered.

Let us consider briefly what would be lost if no

large-scale power plants were operated in any country until world control had been established on such a basis that reasonable confidence could be assured. Chapter IV reviews this situation in detail, and from that discussion we cannot truthfully contend that the immediate use of atomic power is of very great economic importance. In any case, atomic power would first and most logically be used in locations where there are no other sources of energy—for example, in northern Canada, the Amazon River Basin, and other such places where oil and coal are not found. Ships might use atomic power to avoid refueling and the storage of fuel, but this will probably not be economical for some time.

Naval vessels might be more likely consumers of atomic energy than other ships, since the economic factor is not important. However, if we plan to build and operate such naval vessels, we are deciding to prepare for further wars. To be logical we should then decide to make bigger and better atomic bombs, for others will use these bombs if we threaten to use naval vessels or other implements of war against them. If atomic power is developed for naval vessels, the world will inevitably slip back into the atomic-bomb armament race and the whole problem moves back to the use of, and defense against, these bombs. Control of atomic bombs must inevitably lead to the control of all weapons of war. It is foolish to think of controlling only atomic-bomb manufacture and then to let wars start and continue with other weapons until atomic-bomb plants can be put into operation in order to finish the wars. Nothing less than the total abolition of war will prevent their use. In particular the use of atomic power for naval vessels would complicate the control of the military use of atomic energy and, it is hardly necessary to add, almost certainly prevent that degree of confidence so necessary for control.

The postponement of the use of atomic energy for large power plants would make for easier control of the atomic bomb. It is a small price to pay for the accomplishment of this most desirable end. It would not be necessary to postpone the use of radioactive materials at the same time.

Freedom versus an Atomic Armament Race. The citizens of the United States are justly proud of the personal freedom they enjoy. It has been celebrated in song and speech from the beginning of the republic. It is referred to in the first sentence of our Declaration of Independence. It has been continuously lauded in public addresses throughout the land on all public occasions. It has always been imperfect, with violations of its principles at some time in the lives of most individuals and, in some sections of the population, throughout their lives. But our personal freedom, considering the number of people in the country and the length of time over which it has existed, represents a very large fraction of all such freedom that has existed in the whole history of mankind.

There are many circumstances that have led to this condition. The traditions of the early pioneers who settled the country should be remembered. Politically freedom began with the English Magna Charta. But in this country people have attained freedom mostly sooner and to a greater extent than have their blood brothers of Europe. In very large degree this has been due to safety from external aggression provided by the broad expanse of the Atlantic Ocean, behind which we have been able to solve our internal problems without being crushed by a foreign power even during a long and disabling Civil War.

With improvements in transportation and particularly with the discovery and development of air transportation, isolation due to our water defense has disappeared. Today it no longer exists. Twice in this century this country has believed that it was forced to defend itself and its vital interests by sending its sons to fight in Europe and, in this past war, in Asia as well. The same has been true of other similarly situated countries, such as Canada, Australia, and New Zealand.

With the advent of modern airplanes, and now atomic bombs, all natural defenses of all countries of the world have disappeared. Rivers, mountains, and oceans are of no value as defenses now and never will be again. With the obliteration of these defenses, the freedom of this country will be seriously threatened, and, in fact, the threat has already begun. From our beginnings we have known the extent of our armed forces. Today we and our elected representatives do not know the extent of these forces. Atomic bombs are being manufactured in an amount unknown to us, and these bombs represent an armament equivalent to a Navy, a large Army, or a large Air Force. Even Congress does not know the extent of this power, which represents a threat to other countries and has an important effect on the relations of this country to other countries.

If an atomic armament race continues—it is already going on—the citizens of the country will know less and less in regard to vital questions of this kind and finally must accept decisions in regard to public affairs blindly and from a few men in power. Not knowing the size of its armament, the people of the country must trust men in Washington with important decisions previously made through their elected representatives. Men on horseback will rapidly appear on the public scene. Note the attempt to secure the passage of the May-Johnson Bill without proper hearings in Congress. Here was a bill originating in the War Department, which proposed to transfer all control over atomic energy to a few men who would be safeguarded in their acts from all scrutiny by the public through security provisions backed by the most drastic penalties. If that bill or any similar bill passes Congress and is signed by the President, the first abdication of the sovereign rights of the people of the United States will have occurred. The May-Johnson Bill was actually similar in intent and effect to the transfer of power from the German Reichstag to Hitler, though, of course, it would not have so completely destroyed representative government in one act. Many people did not realize the broad and tragic meaning of this bill. It was a definite beginning of the end of our representative government and of the Bill of Rights of our Constitution.

Why have these things occurred? The answer is fear. The atomic bomb is such a grave threat to all men in all countries that frantic and desperate means of handling the problem have been proposed. If the armament race continues, more and more such things will occur. We become afraid, and we destroy the freedom of science. We fear other countries and conceal the number of our atomic bombs. We fear that bombs will be smuggled into our cities and that we will have to introduce secret police to detect such bombs. We fear attacks on our cities and may disperse them regardless of the desires of the people of those cities and of the people of the countryside to which they are moved. We fear sudden attack from without and will transfer the right to declare war from Congress to a single man, and that man, whoever he may be, will be affected by that power. He will become a dictator. Absolute power corrupts absolutely.

The same trend will occur in all countries of the world, and the end will be deadly fear everywhere. But of all the countries of the world, the most industrialized countries will be the most vulnerable and the most likely to be attacked by atomic bombs. These weapons stopped the Second World War, and at the same time they ended the defenses of the United States. They also threaten our liberties.

But why shudder before these fears? Should we not vigorously grasp the situation and take the offensive? The United States might ally with itself as many countries of the world as possible and lead them to conquest of the remainder. In such an undertaking this country would have to supply most of the men and materials. It is a very great effort that would be required, with much sacrifice. Assuming the will to do it, and ultimate success, this nation would become the most hated country on earth, and the hatred would last for a century and perhaps more. Throughout all that time constant vigilance would be required to prevent rebellion in conquered lands. Our people would become brutalized, as the conquering always have. It is not a pleasant solution, and, our traditions being what they are, it would be impossible to secure the will and determination required of our people to carry the program through to the end. Though all proposals for the solution of the atomic bomb are difficult, I believe that this is the most impossible of all.

The advent of the atomic bomb has caused endless confusion in the thinking of men, and the confusion spreads to more people as they come to realize all the implications of this weapon. What do the facts add up to?

The people of the world have in their hands a weapon of transcending size and destructiveness. The knowledge of the existence of this weapon and the methods of its production can never be lost. It can never again be returned to the realm of the unknown. Bombs can be made in large numbers —and cheaply. There is no defense against them. They can destroy physically beyond our ability to comprehend. Fear of them will destroy our liberties. To take the offensive and attempt to dominate the world would wreck our whole lives and those of generations to come.

Civilizations have risen and fallen repeatedly

in the history of the world. We all recall such examples as the Babylonian Empire, the ancient Egyptian civilization, the Roman Empire; and, on this hemisphere, the empires of the Incas and the Mayans. It is to be expected that the future will see rises and falls, too. Modern technological war as developed by the European civilization of which we are a part may cause its complete disintegration. A world war in which atomic weapons are used might very well weaken all of our countries and peoples to such an extent that they would not be able to survive in the future. And not only may our own culture be destroyed by these weapons of mass destruction, but all civilizations as they exist in the world may be retarded and weakened for centuries to come.

It all adds up to the most dangerous situation that humanity has ever faced in all history.

Note

The authors of the different chapters in the first part of this book have tried to set forth the facts about the atomic bomb and to describe the grave situation in which these facts have placed us. How can the menace of the bomb be removed? This is a question not only for physical scientists but for all the people of the world. The statesmen, the experts in international affairs, in government, in political economy, in all the social sciences must speak out, and their proposals must be discussed and weighed in a great public debate.

The problem of how to avert atomic disaster is still very new. We can hardly expect that more than tentative solutions can be brought forth at this time, or that even these will be detailed and concrete. In the following chapters some general approaches to the problem are presented. They will be found to differ with each other and on some points to be diametrically opposed. But they can serve as starting points for fruitful discussion and debate. The points of view set forth are not necessarily subscribed to by the other authors. It is fair to say, however, that all the authors are united in desiring "one world" and that all urge fair and careful consideration of the proposals made here for achieving it.—THE EDITORS.

Leo Szilard, *born in Budapest, worked in Germany until Hitler's rise, then in England (Oxford). In 1939 at Columbia University his experiments became fundamental to the uranium project, and his foresight was largely responsible for governmental support of the project. He has been at the Metallurgical Laboratory, Chicago, since 1941.*

Chapter 12

Can We Avert an Arms Race by an Inspection System?

by LEO SZILARD

CONFLICTS in interest between great powers can be expected to arise in the future as they have arisen in the past, and there is no world authority in existence that can adjudicate the case and enforce the decision if the powers are unable to settle their differences. In the absence of a world authority, conflicting interests could perhaps still be adjusted on an equitable basis by direct negotiation if there were universally accepted principles of law and justice to which the parties could appeal. But as yet there is no such universal acceptance of general principles. Instead negotiations take place in the shadow of the military might that the great powers can muster. In this situation the great powers are inevitably driven to power politics and so long as such is the state of the world the danger of war will exist.

Against this background the existence of atomic bombs creates a new hazard for war. If two countries—and let the two most powerful, the United States and Russia, serve as examples—accumulate large stockpiles of atomic bombs, war is likely to break out, even though neither country has wanted to go to war.

How far can we go towards averting the danger of such an arms race under present conditions—that is, without assuming changes in the general organization of peace that we now have under the United Nations Organization?

If the United States and Russia were to agree to an arrangement ruling out both stockpiles and manufacture of atomic bombs within the territory of either country, it appears very likely that such an arrangement would be acceptable to all other major powers of the world and could be extended to them, or at least to all nations whose voluntary collaboration would be necessary.

If the United States, Russia, and other nations actually set up such an arrangement, an atomic arms race could be postponed and probably averted, provided that it is possible to rule out secret violations. Until there is a world authority capable of enforcing observance among the great powers, it will probably be just as well to let the powers retain the legal right to abrogate their arrangement at any time.

The arrangement itself should provide for rights of inspection to be exercised by an international agency attached to the United Nations Organization. There are a number of ways in which inspection could be made effective, and, while none of the methods may be infallible, all the methods applied together could make violations a very hazardous undertaking.

Inspection of Ores. Aerial surveys, which during the war proved to be very effective, would go a long way toward revealing the presence of mining activities as well as other undisclosed industrial activities. Once uranium mining operations were located, it would be possible to keep track of the mined ores and to follow the uranium from the mine to its destination. If the uranium were obtained from a low-grade ore, mining operations could be detected from the air with a high degree of probability. Nor could the operations be easily camouflaged against infrared photography.

The mining of high-grade uranium ore, in the event that such deposits were discovered, might be somewhat easier to conceal because of the smaller quantity of ore that would have to be mined. But if this mining were carried out in remote and sparsely populated areas, it could still be detected by means of aerial surveys, even though the quantity of ore involved were small. The international agency under whose auspices this survey would be carried out would have to possess the right to issue warrants for searches; then, if necessary, inspectors armed with such warrants could check on the ground any suspicious activities detected from the air.

Mining operations in populous areas, on the other hand, would hardly escape the attention of those who lived and worked in the areas and would therefore scarcely remain a secret for any length of time.

A general geological survey of the world's uranium deposits—which ought to be extended to deposits containing only 1-10 to 1-100 per cent of uranium—would enable us to determine in detail just what measures to adopt for adequate inspection of the mining of uranium in the various parts of the world.

Inspection of Industrial Installations. The detection of secret plants producing U-235 or plutonium presents little difficulty. Plants producing U-235 require such a large supply of power (in the form of either coal, oil, or electricity) that their location is betrayed, particularly if production is concentrated in not highly industrialized regions. If they are dispersed in more densely populated regions, their existence will be known to large numbers of people and will therefore not remain concealed for long.

Because of the heat liberated in the process, plutonium-producing plants can be detected either by the water supply which must be available for cooling, or by some alternate cooling method which would make them easily discernible because of certain peculiar structures involved.

The discovery of any of these plants would be easy during the period of construction. It would be particularly easy within the next few years, since early developments in this field are characterized by more conspicuous installations than those that may follow later.

Inspection of Specialized Personnel. We have so far discussed only more or less mechanical methods of inspection. The over-all aim of preventing an arms race requires, however, that we check not only the manufacture of atomic bombs but also other methods of aggressive warfare, some of which are potentially almost as terrible as those based on the liberation of atomic energy. Such an over-all check, particularly if it is supposed to extend to unforeseen techniques of mass extermination, calls for novel, less mechanical methods of inspection. Knowledge of the movements and activities of all scientists, engineers, and technically skilled personnel would permit the detection of any dangerous activity as soon as it reaches the stage of construction and before it could reach the stage of production. This would be the primary aim of the inspection of personnel.

The inspecting agents must, of course, have scientific knowledge. But college graduates with a fair knowledge of science or engineering, supplemented by a training course of perhaps a few months' duration in certain special fields of knowledge and in the inspection methods they would have to apply, could do the job. They would have to acquire during their college years command of the language of the country to which they go later as inspectors.

Each of these inspectors would have to keep in constant touch with about thirty scientists and engineers in the area assigned to him. If any one of them wanted to conceal some specific fact he could, of course, do so; but it would be very difficult for him to conceal the fact that he was concealing something. A highly industrialized country with as many as 100,000 scientists and engineers who could be used for "high-class" war work would, under these assumptions, require about 3,000 resident agents of the international agency at any one time. Considering the world as a whole and assuming that the average lifetime of an engineer in his profession is about thirty years, it would take just about one year's crop of college graduates in engineering the world over, serving for one year as inspectors, to supply one inspector for every thirty inspected persons. Keeping an up-to-date register for scientists and engineers would naturally be a part of the orderly administration of an inspection service of this type.

Many college graduates might welcome the opportunity to serve for a year after graduation as inspectors in some foreign country, to broaden their knowledge and gather experience in a technical field of their own choosing. Clearly the psychological conditions for successful inspection of this type would be greatly improved if the inspectors were more than police agents—that is, if their function were to spread as well as to receive information. Many of them could engage in a moderate amount of teaching. They could teach their own language or certain specialized subjects in which their own country was more advanced than the host country and convey some knowledge of their native country to their students, many of whom might have to serve there later as inspectors.

The Citizen as Inspector. So far as the great powers are concerned, the problem with which we are faced will be the more easily solved the less we think of inspection in the narrow sense of the term.

In dealing with the threat of the atomic bomb, we are dealing with the unprecedented. Atomic bombs are the product of human imagination applied to the behavior of inanimate matter, and we cannot cope with the problems that their existence created unless we are willing to apply our imagination to the problems of human behavior. These attempts to solve our problems may strike us at first sight as odd, just because they are unprecedented. At this juncture, we must be willing to experiment with problems of human relationships involved in the problem of inspection.

In order to discuss a concrete and well-defined problem, we may again single out the United States and Russia and discuss the possibility of regarding native scientists and engineers of the two countries, rather than foreign inspectors, as the chief guardians of the arrangement.

Scientists and engineers are not isolated from the community in which they live. They have the same loyalties as other members of the community, and their first loyalty may well be to their own country. Just how that loyalty is interpreted will vary, however, with the circumstances. Let us assume that the United States and Russia have arrived at an arrangement which prohibits the manufacture of atomic bombs but which leaves both countries the right to abrogate the arrangement at any time. Let us further assume that after this arrangement has been ratified and become the law of the land, the President of the United States calls upon all scientists and engineers in this country, asking them to pledge themselves to report to an international agency any secret violations committed on the territory of the United States. Let us assume further that the Espionage Act has been modified so that it no longer covers information of a purely scientific or technical nature, whether or not it might relate to the national defense. In circumstances like these there is little doubt that most scientists and engineers in the United States would respond to the President's appeal.

Can we expect Russian scientists to respond similarly? My knowledge of Russian scientists is very much less direct, and my answer to this question must therefore be based on the fundamental conviction that differences between men in general, and scientists in particular, are matters of degree. I do not believe that there are essential differences between Russian and American scientists.

Here it may be desirable to define more closely the conditions under which such a system can be expected to work to command the confidence of all nations. Clearly it would be greatly strengthened by creating international institutions that would establish close collaboration between the scientists and engineers of different countries. The field of atomic energy would be just one of those fields in which large-scale enterprises based on collaboration could be established. Within the framework of such collaboration it would be possible to arrange for every scientist or engineer to spend, in the course of his work, some time during the year outside of his native country, with his family along.

Institutions of this sort could serve a double purpose. First, they would keep alive in the scientists and engineers the already existing higher loyalties shared by all educated men, which transcend the narrowly interpreted loyalties to one's own nation. Second, the frequent and regular occasions at which scientists and engineers would find themselves outside the jurisdiction of their own nation would provide them with an opportunity to report their own government's secret violations of the arrangement to the appropriate international authority without endangering their lives or the safety of their families. They could be effectively guaranteed immunity, assuming they were willing to remain outside the jurisdiction of their native country. If they did so, they would have to be guaranteed the right to choose their residence abroad, and an appropriate source of income would have to be provided for them.

Naturally, no scientist or engineer would find it an easy decision to become an exile. But after all, under a sensible inspection system, secret violations of the arrangement would have to be regarded as very unlikely occurences, something like a major catastrophe on the road leading to abrogation, to an arms race, and to war. Viewing it in this light, the scientists and engineers would be inclined to look upon the necessity of reporting a secret violation on the part of their own nation as a personal misfortune—a misfortune small compared to the disaster that the violation of the arrangement itself would forebode for the world.

The fact that scientists and engineers would be in a position to report violations without risking their lives would help to alleviate suspicion that they knew of secret violations but were keeping silent for fear of their lives. However exaggerated this kind of suspicion might be, it could become dangerous in times of political stress, inasmuch as it might lead to a bona fide abrogation of the arrangement by one of the major powers.

A country thinking in terms of power politics might be tempted to abrogate, or to threaten to abrogate if by doing so it could greatly shift the balance of power in its own favor. This would be the temptation either if it could quickly outproduce its potential enemy or if it were much less vulnerable to bombs than its potential enemy. The desire to abrogate from this motive would be weaker if it would take a long time from abrogation until bombs could be made available in substantial quantities. That time might be anywhere from six months to three years, depending on whether atomic power installations for peacetime purposes were in existence at the time of abrogation and depending upon the restrictions that might have been imposed upon these installations. Renouncing for the next ten to fifteen years any large-scale use of atomic energy for purposes of producing electrical power might, therefore, tend to remove incentives to abrogation. This would be a very much smaller sacrifice for the United States than for other countries that are in greater need of electrical power and much poorer in natural resources.

Need for a Long-range Program. We cannot expect, however, to hold up indefinitely the peacetime uses of atomic power for the sake of security, and we shall have to go as soon as possible beyond such temporary expedients.

An arrangement of the type that we have discussed would remove the threat of an arms race and would be of great value because under it war would break out only if one of the major powers actually decided to risk a war by abrogation. If we eliminate not only atomic bombs but also the development of other aggressive methods of warfare, and particularly eliminate stocks of long-range aggressive weapons, such as long-range bombers, large fleets of warships, and landing craft, the risk of war between the great powers will appear to be remote and a tolerably well-working peace system under the United Nations Organization, as at present constituted, might be expected to function for a while. We cannot hope, however, to safeguard peace forever under such an arrangement.

We may have removed for the time being the danger of one kind of war—the war that arises more or less automatically out of an armed peace in which the great powers maneuver according to the laws of power politics. The First World War may perhaps be cited as an example of this kind of war which could be averted under such an arrangement, but the Second World War, in which Germany deliberately set out to conquer, does not fall into this class. Under the arrangement discussed in this chapter, there would remain a definite danger in any one year, of war's breaking out.

The breathing spell that we might secure by averting an arms race would give us the opportunity to establish a world community. Unless we made use of it for this purpose, we would have done nothing but postpone the next world war, which will be all the more terrible the later it comes. The issue that we have to face is not whether we can create a world government before this century is over. That appears to be very likely. The issue that we have to face is whether we can have such a world government without going through a third world war. What matters is to create at once conditions in which the ultimate establishment of a world government will appear as inevitable to most men as war appears inevitable at present to many.

Clearly the crucial point in this transition will be reached when a world government will in fact operate in the area of security or police functions. When that point is reached, the right to abrogate will cease and secession will become both illegal and in fact impossible.

Discussion of such a long-range program would

go beyond the scope of this chapter. I have mentioned it because I doubt that the danger of an arms race can be successfully averted unless the problem of creating a breathing spell and the problem of establishing a world community—that is, the short-range and the long-range programs—are attacked simultaneously. For, if we wish to avert an arms race, we will have to give up our own atomic bombs and scrap our own manufacturing facilities before we can have a foolproof peace system. We shall have to take risks, and we shall have to derive the courage to take risks from the conviction that we are on our way toward the solution of the problem of permanent peace.

WALTER LIPPMANN, *the author of many books on political affairs and for 15 years a special writer for the New York* HERALD TRIBUNE, *has become known through his writings as one of the foremost political commentators of the day.*

Chapter 13

International Control of Atomic Energy

By WALTER LIPPMANN

LET us now examine the problem, as it was defined by the three foreign ministers, of how to achieve "control of atomic energy . . . for peaceful purposes."*

My task is to inquire into the prospects of solving this problem, given our present knowledge of politics, government, and law. For at the outset we have to recognize that our progress in the art of mass destruction has not been accompanied by new discoveries in political science or in statecraft. We have not learned how to release hitherto inaccessible intellectual and moral energies and to direct them to constructive ends. To start with we have only the political science of the preatomic age. And while we may assume that the terrifying character of modern total war will make men somewhat more willing to support political inquiry and experiment, nothing can now be proposed that is not an application of knowledge that already exists.

Nevertheless I shall contend, and I hope to demonstrate, that the political principles of the solution are known. Whether mankind in our generation will apply them is another question. It is of the utmost importance, to be sure, but it is a question that we cannot begin to examine until we have elucidated the theory of the solution. For the practical difficulty, which is how to persuade men to accept a solution, cannot be approached until we see clearly what it is that they must be persuaded to accept.

All will agree, I believe, that the crux of the immediate problem is how to provide "for effective safeguards by way of inspection and other means to protect complying states against the hazards of violations and evasions."† For as Secretary Byrnes said on his return from Moscow:

> In particular it was intended and is understood that the matter of safeguards will apply to the recommendation of the commission (to be established by the General Assembly of the United Nations) in relation to every phase of the subject and at every stage. Indeed, at the root of the whole matter lies the problem of providing the necessary safeguards.

It is evident that international rules will be only as good as the safeguards against their violation and evasion are effective. The fundamental problem, in short, is how to enforce the international agreements that the governments may decide to sign. For agreements are not likely to be observed if men do not have reason to believe that they will be enforced. The stakes are the life and death of national states and of masses of their inhabitants: No nation could afford the risk of being a complying state unless all states capable of producing these weapons were assuredly complying states as well.

Declarations and resolutions that are unrelated to the means for enforcing them may, and often do, serve a great purpose. They may edify, teach, inspire, and illuminate the possibilities of the future. But they are not law—even if everyone has subscribed to them—and here and now we are concerned with the making of international agreements that will have the force and effect of world law. The prospects of enforcement are the controlling considerations: we can draw up only such rules as we have reason to believe we can enforce. Indeed, the very question of whether there shall be an international policy at all, rather than a national one alone, depends on what faith and credit we can put in the enforcement of international agreements.

There are few in any country who now believe that war itself or any of the important weapons of war can be regulated or outlawed by the ordinary treaties among sovereign states. During the years between 1919 and 1939 many treaties were signed

* Communique on the Moscow Conference of the Three Foreign Ministers, December 27, 1945.

† *Id.*, VII, V, D.

and ratified.* The sovereign states promised to maintain peace, to outlaw war as an instrument of national policy, to limit their armaments; they gave and received bilateral and multilateral guaranties of mutual protection and nonaggression. These treaties did not prevent nor did they mitigate the fury and the horror of the Second World War of the twentieth century. They were not observed by the aggressor states nor enforced successfully by the complying states. No reliance can now be placed, nor indeed will it be, on more treaties of the same kind. No matter how solemn the language of the new treaties, or how specific and comprehensive the substance and the procedure, no one will put his trust in them.

But it is just as evident that there is no way to *begin* to deal with our problem except by international treaties of some sort. We should merely be begging the question if we did not recognize that no world-wide proposal can be adopted except by a treaty that the sovereign states of our epoch will ratify. We have, therefore, to inquire into the exact reasons why treaties of the old sort are defective; if the diagnosis is correct, it should lead us to the remedy.

As we have known them in our time, nearly all international agreements and almost all international law have been enforceable only in so far as sovereign states would and could coerce other sovereign states. They have provided no other method of enforcement except that the complying states should in the end be ready and willing to wage war against transgressor states. This has been true, as the words themselves indicate, even of measures that have been called "short of war"—of diplomatic nonintercourse, embargo, and blockade. The milder penalties were counted upon to be deterrent only because each measure short of war was to be less and less short of war. They were regarded as a series of measures that might begin with the withdrawal of an ambassador and could end in total war. The effectiveness of any sanction depends upon the fact that it is a warning and token of severer penalties to come. The world has seen this in the case of Japan in Manchuria, of Italy in Abyssinia, of Germany in Austria, of Spain and Argentina during the Second World War: the initial sanctions were not deterrent because the complying states were not ready or willing to apply the final sanction of going to war.

The enforcement of international agreement by sovereign states against sovereign states is known as the method of collective security. We cannot rely—indeed no nation does or will rely—upon international agreements of this kind. Why not? Because the remedy is as bad as the disease: the peaceable nations have to be willing to wage total war in order to prevent total war. The remedy is so crude, so expensive, and usually so repulsive, that it will not be applied by the very peoples who are supposed to apply it, namely by the peace-loving peoples.

We must be clear about this, for much hangs upon it. It is often said that the mere threat of collective force will deter any state from taking the steps that lead to war. That might be true if the threat is known to be genuine—if it is not a gesture and a bluff. There must be no doubt in the minds of the rulers of the transgressor states that the others are mobilized, equipped, and trained, and it must be certain that there will be no hesitation and debate about the willingness of the law-abiding peoples to wage total war. To state these conditions is to know how improbable it is that they will be met in times of peace. For in the early stages of any campaign of conquest, the issues are certain to be remote and in themselves of no great importance to the nation that must carry the main burden of collective security. We may recall the seizure of Manchuria in 1931–1932, Ethiopia in 1935, the Spanish civil war in 1936, the reoccupation of the Rhineland in 1936, the episode of the Panay in 1937. It is in these early stages of aggression that collective security would have to be effective if war is to be prevented. But that is just when it is least effective: the peace-loving states cannot be counted upon to be ready and willing to wage total war over what appear to be in themselves minor, remote, and unclear disputes. Their unreadiness and unwillingness will be patent to the aggressor, and therefore their collective threats will be discounted as a collective bluff.

The threat of total collective war can be a deterrent only if it is evident that the threat will be car-

* For example, the Covenant of the League of Nations; the Washington Conference Treaties Limiting Naval Armaments, Relating to the Use of Submarines and Noxious Gases in Warfare; the Nine Power Treaty on the Far East in 1922; the Locarno Treaties of 1925; the series of Conventions of Arbitration and Conciliation between Germany and the Netherlands in 1926, Denmark, 1926, and Luxembourg, 1929; the Kellogg-Briand Pact, 1928; German-Polish Nonaggression Pact, 1934; Austro-German Agreement, 1936; Munich Agreement, 1938, Nonaggression Treaty between Germany and Denmark, 1939; and between Germany and U.S.S.R., 1939.

ried out. But it will be carried out only as the very last resort. It is in fact not a way of enforcing international agreements. It is a measure of ultimate desperation that will be used only when reliance upon agreements is gone, the peace of the world has already been irretrievably shattered, and the peace-loving nations are compelled to unite in order to fight a war of survival.

When the issue is less than the survival of the great nations, the method of collective security will not be used because it is just as terrifying to the policeman as it is to the lawbreakers. It punishes the law-enforcing states, at least until they have paid the awful price of victory, as much as the law-breaking states. Therefore, it cannot be used as a method of ordinary and continuing enforcement, for example, as a means of insuring the inspection of laboratories and plants working with fissionable materials. There would be little surgery if the surgeon had to amputate his own arm when he was called upon to amputate his patient's leg. There would be little enforcement of law in our cities if in order to arrest burglars, murderers, and violators of the traffic ordinances the police had to start a fight in which the courthouse, the jail, and their own homes were likely to be demolished. Men will not burn down the barn in order to roast a pig: the method of collective security is, I repeat, too crude, too expensive, and too unreliable for general and regular use.

It proposes to achieve peace through law by calling upon great masses of innocent people to stand ready to exterminate great masses of innocent people. No world order can be founded upon such a principle. It cannot command the support of civilized men, least of all of democratic men who respect the individual and consider it the very essence of justice to distinguish between the guilty and the innocent, the responsible and the irresponsible.

Our own experience with the method of collective security has proved how right was Hamilton in saying that when "every breach of the laws must involve a state of war and military execution must become the only instrument of civil obedience," no "prudent man (would) choose to commit his happiness to it."*

At the beginning of this chapter I said that the essential political principle is known by which our problem can be solved. There is no mystery about it, and indeed it becomes self-evident once we realize clearly why collective security is such a bad method of enforcing laws and agreements. The principle is to make individuals, not sovereign states, the objects of the international agreements; it is to have laws operate upon individuals.

This principle is not altogether novel even in the international affairs of our era when national sovereignty has been so absolute, and its doctrines have been expounded so dogmatically and so pedantically.† It is the principle that men have had to invoke and apply "whenever they have sought to enlarge the area of lawful order."‡

The authors of the American Constitution invoked it in order to remedy the lawlessness and disorder of the Confederacy of 1781. They espoused and elucidated the principle in the Federalist.# If anything in the field of political science can be called a proven discovery, it is that a system of law will not produce order if it operates only upon states, and that the enforcement of law becomes possible only as the laws operate upon individuals. For then the enforcement of the law may not encounter "the organized and unified opposition which is evoked" when the attempt is made to regulate or coerce states that command the allegiance and obedience of masses of people.

Hamilton argues that if there is to be a "superintending power"—which is what we are committed to establishing when we seek "effective safeguards" against weapons of mass destruction—then "we must resolve to incorporate into our plan those ingredients which may be considered as forming the characteristic difference between a league and a government; we must extend the authority of the Union (in this case of the superintending power of the United Nations) to the persons of the citizens" of the United Nations.

In examining the bearing of this principle upon

---

* Federalist Papers, No.. 15.

† Cf. Hans Kelsen, *Peace Through Law*, University of North Carolina Press, pp. 71 *et seq.*, for instances of individual responsibility established by general international law or treaty—namely, rules forbidding piracy, breach of blockade and contraband, illegitimate warfare; also Article III of the abortive Treaty of Washington, 1922, on submarine warfare and Article II of the International Convention for the Protection of Submarine Telegraph Cables, 1884.

Other interesting and suggestive instances are: Treaty for the Suppression of the African Slave Trade, 1862; International Convention for the Suppression of the Traffic in Women and Children, 1921; International Convention for the Suppression of the Circulation of and Traffic in Obscene Publications, 1923; and the International Convention for the Suppression of Counterfeiting Currency, 1929.

‡ Report of the Committee on International Law of the Association of the Bar of the city of New York, June, 1944 (John Foster Dulles, chairman).

Nos. 15-20 and No. 27.

the problems of the world today, we must not be diverted or confused by the connotations of the word "government." The word suggests the apparatus of a world flag, a world executive, a world legislature, a world judicial system, a world army, world policemen, detectives, inspectors, and tax collectors. None, some, or all of these instruments of government may be desirable or feasible; the point I wish to insist upon is that we need not and that we should not consider them now. For the principle that world laws and agreements shall operate upon individuals can be applied constructively at once without a priori commitment to create the particular institutions of a world government.

The principle is most suitable to the problem that the three foreign ministers agreed to lay before the Commission for the Control of Atomic Energy that they have asked the General Assembly of the United Nations to establish. The problem is how to provide "effective safeguards . . . against the hazards of violations and evasions" of agreements that would call for "the exchange of basic scientific information for peaceful ends," for the "control of atomic energy to the extent necessary to insure its use only for peaceful purposes," and "for the elimination from national armaments of atomic weapons and other major weapons of mass destruction."

It is manifest that these rules will deal with the activities of countless individuals in all countries. Scientists, technicians, industrialists, administrative officials, inspectors, judges, legislators, military commanders, diplomats, and the rulers of states must comply with the rules. They must enforce the rules. They must be accountable for violating or evading them. They must be protected against being forced to violate or evade them.

If mankind is to rely upon obedience to the law by such a multitude of individuals, the rules agreed upon must become the supreme law in all lands, and all previous and subsequent national law must conform to the world law. A nation which refuses to accept this had better not be invited to sign the treaty. For it will be pretending to subscribe to rules that its own laws do not support. Thus, by invoking this principle we can establish at the outset a clear criterion as to whether there is in fact any good prospect that the safeguards will be effective. We can stipulate that no state shall be held to have ratified the treaty until by domestic legislation it has expressly made the rules of the treaty the national law within its jurisdiction.

But that is not all. Since the treaty would require that the laws governing atomic energy be essentially the same throughout the world, the United Nations could hold that any individual person was entitled to the protection of that law and was liable under it in any jurisdiction of any of the member states. Then no one who violated the law could claim the protection of his own government. He would be an outlaw, like a pirate, who could be indicted, arrested, tried and punished in any of the United Nations. If his defense was that he had acted under the order of superior officials of his own government, it would not be an unfriendly act but an established right to ask that government for explanation and investigation. If the government refused, then, of course, it would be in rebellion against the United Nations, and the hard question—which can arise in any civil society —would be posed as to whether they would resort to war to suppress the rebellion. If it came to that, the rulers of the rebellious state would be liable to indictment as war criminals and, if ever they were caught, to trial and punishment.

Any individual scientist, industrialist, administrator, or official who wished to obey the law could, if his government were seeking to coerce him, claim the protection of the United Nations. If he escaped, they would give him asylum. If he were put in a concentration camp, the United Nations could demand an explanation and a fair hearing of his case if any friend or relative managed to convey the news of his case to any agent of any government of the United Nations.

No one would owe allegiance to his own state when that meant that he had to violate the world law. It would not be unpatriotic, in fact quite the contrary, for any man to expose officials who were conspiring to violate what would be the law of the world and the law of their own country. He could expose them with a good conscience just as he would expose them if in the United States they were conspiring against the Bill of Rights, or for that matter to rob the Treasury. They would be the traitors, the usurpers, the disloyalists, the criminals, and lawbreakers. He would be the law-abiding citizen of his country and of the world, and, if he took risks in order to uphold the law, he would have behind him the power of all law-abiding states and the explicit and avowed conscience of mankind.

While this principle can be applied progressively to enlarge the area of order under world law, it is especially suited to the specific problem of how to provide effective safeguards against violation and evasion of international agreements about atomic energy. These agreements have not been worked out as this chapter goes to the printer. But their general character and purpose have been sufficiently forecast by the Truman-Atlee-King Declaration of November 15, 1945, and the Moscow communiqué, and I am assuming that our main concern is with how they are to be observed and enforced.

The proposed agreements will be designed to limit the development and use of atomic energy to peaceable ends and purposes. This must mean that at no stage in the process from pure research and the mining of ores to the manufacture of weapons can secrecy be permitted which would enable a government or a faction of conspirators to use atomic energy for ends that were prohibited by the agreement. The disclosure and the inspection must be adequate to make it highly improbable that complying states will be made the victims of sinister surprise and sneak attack. Enough must be known so that the complying states can be forewarned in time to take preventive and defensive measures. This need not mean that everyone must be taught how to make atomic bombs in the kitchen sink. But it must mean that no government can even start to prepare itself to make atomic bombs except with the consent of the other governments and in accord with the international rules they have agreed upon.

It follows, therefore, that treaties must be designed directly to nullify the sovereign right and to destroy the actual power of any government to make a state secret of the development of atomic energy. A state secret is kept by means of national laws and regulations establishing censorship, definitions of treason and of espionage, and secrecy is enforced by restrictions upon all who share the secret and penalties upon all who might ferret it out. If, then, members of the United Nations are to agree to the mutual right of inspection, they must agree that in these matters the sovereign right to enforce a state secret is no longer absolute. The apparatus of censorship, treason, and espionage is, as respects the terms agreed upon, null and void.

Even in time of war among well-nigh absolute sovereign states, the enforcement of complete secrecy is exceedingly difficult and in large degree imperfect. The kind of agreement that we are discussing would make it much more difficult, especially in peacetime. It would make secrecy by government officials unlawful and would make it lawful, and also righteous, honorable, and not too imprudent, for anyone to expose violations of the rules and to inform the inspectors.

Dr. Szilard examines the details of the problem of inspection in Chapter 12. It may be added here that, under agreements of the type we are examining, the prohibitions that would prevent effective inspection are outlawed, and therefore the inhibitions of individuals arising out of patriotism or fear of prosecution are greatly reduced. It would cease to be a crime against the state to help the inspectors: it would have become a crime to obstruct them. Individuals who wished to observe and to enforce the world law on this subject would have the support, once they had managed to invoke it, of the combined power and influence of all the complying states. We need not suppose that the complying states will rely wholly upon United Nations inspectors wearing badges to identify them; they will maintain also diplomatic and consular agents, intelligence services, and there will be spread all over the world journalists, businessmen, tourists, missionaries, and students. It would still be theoretically possible, but it would be much more difficult, for another Hitler to lock up an anti-Hitlerite or to have him disappear surreptitiously, without some word of it being sent by the man's family or friends to some agent or even a mere citizen of a complying state.

There is every reason to think that the international family of scientific men would become the foremost supporters of the international agreements we are discussing. These agreements would recognize, legalize, and protect the established traditions of scientific men: the agreements would authorize them, invite them, and induce them, to do the very things that they need to do and must want to do. Because atomic energy cannot be developed without them, they occupy a strategically controlling position. They are, therefore, the naturally appointed guardians of any system of international control. They would be most expertly qualified to draw conclusions from the reports that would come in not only from the formal inspectors but from all other services of intelligence and information.

Our agreements will be sound law not only because their purpose is good but because they enable the many scientists and technicians to serve their own interests and professional ideals. It is easier to administer laws like these which release multitudes of men than laws which restrict them. Agreements of this type would use the liberty of the individual to regulate the absolutism of the national state.

All this will become possible if we found the treaties we propose to ratify upon the basic principle that they prescribe rights and duties not only for states but for individuals. But if we do not introduce this ingredient, as Hamilton called it, then the agreements will not be laws. They will be declarations only. For observance will depend upon the faithful performance by all sovereign states, and enforcement upon the willingness and readiness of some states to wage total war in the name of collective security.

Yet our conclusions, however cogent, would have no practical importance if they were merely a theoretical demonstration that atomic weapons can best be regulated in a certain way. We should then have just another plan for the limitation of armaments, and we know from the experience of 1919 to 1939 that partial disarmament does not prevent war and may indeed prove to be a snare and a delusion for those very nations that put their trust in it. In the end we must be concerned not with atomic war but with war, since we know perfectly well that the best possible system for regulating bombs will be swept away if there is another great war. If there is another war of the giant powers, atomic and even more deadly and malignant weapons* will—we must assume—be used. For even if they are not in stock when the war breaks out, they will be manufactured before it is concluded.

Therefore, in judging specific plans to control atomic energy, we must examine their bearing upon the formation of a world-wide order of peace. We must see how the principle, which I have been contending is suitable to the control of atomic energy, affects the United Nations as a world society and the United Nations Organization that they have just established. Consistency is essential: there cannot be a system of world law that is unique for atomic energy and a different and conflicting system for the maintenance of peace.

There is, however, no conflict. On the contrary, the method we have been discussing for regulating atomic energy is a concrete application of the fundamental principle to which the United Nations are already committed by implication and by their acts. I realize that many ardent and faithful supporters of the old League and of the new organization think otherwise. Yet it is demonstrable, I think, that the United Nations have in fact rejected the method of collective security, and in so far as they have adopted any method of enforcing agreements and laws, it is to found the international order upon laws that govern individuals.

The charter of the United Nations organization does not explicitly reject the idea that peace is to be kept by authorizing a universal war against transgressor states. The charter does indeed say it is one of the "purposes" of the United Nations to take "collective measures . . . for the suppression of . . . breaches of the peace,"† and it empowers the Security Council‡ "to take such action by air, sea, or land forces as may be necessary to maintain or restore international peace and security."

But, as everyone knows, all this is nullified by the rule of unanimity, usually called the privilege of veto, among the five great powers.# The method of collective security cannot be used lawfully against any one of the great military powers without its consent. This is equivalent, of course, to saying that it can never be used. For no nation will ever conceivably authorize the rest of the world to wage total war against itself. Moreover, the rule of unanimity protects all other states against collective coercion unless, perchance, there is some state so small, so isolated, and so unimportant that it is not the ally or client of any one of the great powers.

Thus the United Nations, when they framed the charter of their organization, renounced in practice though not in theory the method of collective security. There are many who deem this a reactionary event in international affairs and argue that every effort must be made to abolish the veto

* Cf. Navy Department report on research in biological warfare, January 4, 1946.

† Charter of the United Nations, Chapter I, Article 1, Section 1.

‡ *Id.*, Chapter VII, Article 42.

Id., Chapter V, Article 27, Section 3.

and to establish the principle of collective security. They will have, I believe, to reconsider their position. In 1919 the United States rejected the Covenant because it would not submit to or participate in a commitment to wage war to make peace. Neither against Japan in 1931 nor even against Italy in 1936 were the members of the League of Nations willing to fulfill their commitment. In 1945 the Soviet Union most positively, and the United States most probably, would not have ratified the charter if it had in fact authorized the method of collective security.

Since the great powers have in fact rejected collective security, it does not follow that they are international anarchists. The great powers may be right not because they are great powers but because they are so directly responsible and so immediately involved in the consequences that they are compelled to see the true nature of collective security. They may have rejected it not because they are wrong-minded but because the method is wrong—because it is in truth too crude, too expensive, too unreliable, and also too unjust, to be used generally and continually for the enforcement of international conventions.

In any event it is the fact that without a revolutionary revision of the charter the method of collective security cannot be used to control atomic weapons or for any other purpose. Those who argue that the veto must be abolished if there is to be any sanction behind agreements and laws are taking a position that is tantamount to renouncing all hope of an order of law in the world. If there is no other method of enforcement, then there is no method of enforcement, and we find ourselves in a world condemned to the unending anarchy of sovereign states.

But as a matter of fact if we look beyond the San Francisco Charter to the United Nations as a living world society, we see that during the past quarter of a century, as they have rejected the method of collective security, they have also been committing themselves deeply to the other method that we have been discussing, that is, to holding individuals accountable for breaches of the peace and for the violation of treaties and of international law.

The commitment is now solemn, deep, and publicly declared. It is sealed by the fact that all the United Nations have participated in the arrest, the indictment, the trial, and the punishment of war criminals. No one has protested, and by their words and their acts all are committed to the doctrine enunciated by Mr. Justice Jackson in his opening address at the Nuremberg trial that "the forces of law and order be made equal to the task of dealing with such international lawlessness as I have recited here" by taking "the ultimate step," which is "to make statesmen responsible to law." With the assent of his British, Soviet, and French colleagues Mr. Justice Jackson completed the commitment: ". . . and let me make it clear that while this law is first applied against German aggressors, the law includes, and if it is to serve a useful purpose it must condemn, aggression by any other nation, including those which now sit here in judgment."

We shall have misunderstood the real principles that govern the United Nations if we do not see that as they have rejected the principle of collective security by adopting the veto, they have embraced the principle that "crimes are always committed by persons" and that "only sanctions which reach individuals can peacefully and effectively be enforced." The commitments of the Nuremberg trial are no sudden improvisation out of thin air: they have their roots in the history of our epoch, and they have been evolved during the two World Wars. Though they are, like all common law in its beginnings, empiric and uncodified, they are no less authoritative than the Charter. The fundamental law of the United Nations is not confined to the Charter, and in construing the text of the Charter we must take fully into account the law that they have promulgated at Nuremberg.

Here the United Nations have recognized, in the words of Mr. Justice Jackson's opening statement, "individual responsibility on the part of those who commit acts defined as crimes, or who incite others to do so, or who join a common plan with other persons, groups, or organizations to bring about their commission. The principle of individual responsibility for piracy and brigandage, which have long been recognized as crimes punishable under international law, is old and well established. That is what illegal warfare is. This principle of personal liability is a necessary as well as logical one if international law is to render real help to the maintenance of peace. An international law which operates only on states can be enforced only by war because the most practicable method of coercing a state is warfare."

We can now see in retrospect that during the Second World War there was consummated a revolutionary development in human relations. It has pushed mankind across the boundary lines of what was until recently the modern age, when men lived in a congeries of unqualifiedly sovereign states and their dependencies; the first but essential formations of the world state have begun. They are not merely being proposed and advocated. This event was not caused, though its evolution may now be accelerated, by the portent of the atomic bomb. For the decisive change occurred before the explosions at Los Alamos, Hiroshima, and Nagasaki.

Like all great historic events it was not originally the product of a conscious design but of a series of necessary decisions taken for empiric reasons. The United Nations became an alliance because they were all, though separately and at different times, the victims of aggression. They were compelled to unite in order to wage a war of survival. They were moved to make their union permanent by the knowledge that in no other way could they hope to consolidate the peace they had paid such a price to win. But when they came to write the charter of their union at Dumbarton Oaks and at San Francisco, they found it impossible to construct an international order on the principle of collective security. As sovereign states they could not participate in a world order within which sovereign states were authorized and obliged to wage war against sovereign states.

This did not mean, as many men have thought, that the United Nations are stalled on the way to a world order. It meant, on the contrary, an admission on their part that out of sovereign states alone a world order cannot be formed. Though that was not the intention, nor was its significance appreciated when it happened, what was in fact blocked by the rule of unanimity was the effort to advance along a way that did not and could not take the United Nations into a world order of law.

But simultaneously, though separately, they were moved to open up the way that can lead to a world order. The impelling cause here was in the first instance their attempt by warnings and threats to stop the massacres and atrocities that were an integral part, not merely incidents, of the Nazi doctrine and practical conduct of war. These warnings were not heeded. Then the impelling motive of the Allies became the need to inflict retribution upon and to exact some rough measure of justice from those who were most clearly responsible for the monstrous evils of the war. The Allies united in applying the principle that not the anonymous collective entity of the state but the responsible officials of the state may be held personally accountable for the violations of treaties, of the conventions of war, and of the covenants of international law.

With the Nuremberg trial itself, which is not concluded as these lines are written, we are not concerned. Our conclusions are unaffected by the questions that have still to be determined, whether any or all of the defendants are justly and legally guilty on all the counts in the indictment. They could all be innocent, or they could plead successfully that the law under which they are being prosecuted is in their case *ex post facto*. None the less this would be the law of the United Nations henceforth unless we intend to deny the authority of law to what they have all declared, sealed, signed, and ratified, and repeatedly affirmed, by the official action of their lawful governments.

Nor can it be said that this principle of personal liability is a new doctrine and alien to the conscience of civilized men. I believe it could be shown that it is the traditional and orthodox doctrine and that the theory of the absolute sovereign state that is subject to no higher law, and is itself the source of the highest law of its people, is an aberration and heresy, which has flourished, though even then never without protest, during the closing decades of the nineteenth and the opening decades of the twentieth century.

To President Wilson belongs the distinction of having been the first head of a great state to attack the foundations and premises of this heresy. He did just that on April 6, 1917, when in asking Congress to recognize that the United States was at war with the Imperial German government, he said that we should fight not against the German people but against "their rulers." It does not matter how many or how few Germans he or anyone else adjudged to be guilty and responsible: once it was declared that in a war not all the inhabitants of an enemy state are collectively indistinguishable, the doctrine of the absolute sovereign state had been breached.

The principle that Wilson advocated in his message was carried over into the Treaty of Versailles when the Allies and Associated Powers did "publicly arraign William II of Hohenzollern, formerly German Emperor, for a supreme offense against international morality and the sanctity of treaties."

But in 1919 the nations were reluctant as a matter of doctrine and unprepared for reasons of expediency to act on the principle.

In 1945 they did act on it. They did so only after explicit, repeated, and formal declarations that they would act on it. Thus on January 13, 1942, twenty-eight months before the defeat of Germany, an Allied conference of nine occupied countries of Europe placed "among their principal war aims the punishment, through the channels of organized justice, of those guilty and responsible for these crimes, whether they have ordered them, perpetrated them, or in any way participated in them." This declaration had the approval of the United Kingdom, the British Dominions, the U.S.S.R., China, India, and the United States, all of them participating as observers. They themselves subsequently and repeatedly made a similar commitment, and from these declarations stems the prosecution of the war criminals.

While the United Nations had begun empirically, seeking first to deter the enemy officials and "those who have hitherto not imbrued their hands with innocent blood,"* and then to exact retribution, the evolution of their doctrine and practice took them far beyond the case of the criminals of this war. At the Nuremberg process they bound themselves to the general principle that not only these German aggressors but all future aggressors shall be accountable to the same law. By this engagement the United Nations adopted the elements that form "the characteristic difference between a world league and a world state."

As we look back upon what has happened, we see what can happen. The United Nations Organization is not another League of Nations, rendered impotent by the veto. It is the constituent association of a world state already directed to establishing a universal order in which law, designed to maintain the peace, operates upon individual persons.

No one can prove how fast and how far mankind will now go to form the world state, or what will be the legislative, executive, and judicial organs of the world state. We may indeed fail altogether and be doomed to the desolation of utter anarchy. Nothing that could happen will surely happen. But what we can prove—and it is a momentous conclusion—is that the potentiality of the world state is inherent in the United Nations. When I say that it is inherent, I mean that this is the end and the logic according to which the United Nations must evolve if they are to evolve at all towards an enduring world order. The world state is inherent in the United Nations as an oak tree is in an acorn. Not all acorns become oak trees; many fall on stony ground or are devoured by the beasts of the wilderness. But if an acorn matures, it will not become a whale or an orchid. It can become only an oak. That is the potentiality inherent in its organism. In this sense, not another League of Nations but a world state, in the exact meaning of the term, is inherent and potential in the embryonic organism of the United Nations.

The recognition of this truth will be in itself an event that will affect the course of events. For when an idea that enlists men's hopes is seen to be consistent with their acts, it evokes and organizes their energies. It is not then an abstraction or an essence. It is a dynamic force in their conduct. There are the ideas that shake the world and change it.

The project of the world state is now such an idea. It was not always such an idea, though for some two thousand years in the Western world, at least since the Stoic philosophers were teaching, men have been able to transcend their tribal inheritance and to imagine the ideal of the universal state. But for long ages they were able to imagine also many other things they could not achieve—that, for example, men could fly and that no human person should be a chattel slave. Much had to happen, much had to be experienced and discovered, before the abolition of slavery or the art of flying could become realizable ideas. So it has been with the ideal of the union of mankind under universal law. Much has had to happen, much experienced, discovered, and learned, before the leading peoples of the world could arrive at the point where the formation of the world state

* Declaration by President Roosevelt, Prime Minister Churchill, and Premier Stalin, issued at the Tripartite Conference at Moscow, November 1, 1943.

was not only what many of them desired but what in fact they were engaged in creating.

Now the ancient ideal has become an idea, indispensable in fact, to which men have been compelled to turn: there is no other way they can exact justice for the crimes of the war, no other way they can establish effective safeguards against the misuse of the weapons of mass destruction, no other way they can hope to make international agreements enforceable. The solution of the most urgent practical problems and the advance towards a wider and greater order of peace among men depend upon the same fundamental idea. We need not hesitate to recognize it and to proclaim it and to employ it as the creative principle of the coming order of mankind.

Not our powers of persuasion but the inevitability of the truth will convert men's minds and enlist their support. We need not suppose that all the nations and all the peoples in them will become suddenly unanimous and incandescent with enthusiasm to found the world state. Some must be convinced before many can be convinced, and under any predictable conditions the affairs of the world will long be determined by the rivalries, the combinations, and an uneasy equilibrium among the sovereign and powerful states. But in this condition of the world, a new event may and can, and could be made to intervene. That event would be the decision of the American people to make the formation of the world state the principal objective of their own foreign policy.

This can be made to happen. For the American people, who have learned that they cannot live in isolation, have no taste and no aptitude for the career of a world power among world powers, and no faith that good can come of it. Intuitively and by tradition they believe that security and serenity and great achievements require a universal order of equal laws and can never be had in a mere equilibrium of sovereign states. It would conform, therefore, with American ideals and with American interests to dedicate the power and influence of the United States to sponsoring and pressing for the formation of a world state.

If that were the dynamic core of the foreign policy of the United States, the influence of this decision upon mankind would be enormous. The United States is at the zenith of its power and for the time being at least is the sole possessor of the most devastating weapons ever manufactured on earth. There is no doubt that if at this moment in history the United States raised the standard, many nations would immediately rally to it and in all the other nations more and more of the people.

Never was there such an opportunity for any people as is ours, though briefly if we do not seize it. We can use the preeminence of our military power so that an ideal for all mankind, not the United States of America as a national state, may dominate and conquer the world. The issues of territory and of resources and all the other knotty problems of settling the war and of making peace will still have to be dealt with. The struggle of civilized men with the primitive, the stubborn, the malign, and the stupid, within each of us and all about us, will not end. But how different would be the assumptions and the expectations of diplomacy if, as a great power, in the company of other nations who would surely be with us, we were committed to the formation of a world order of universal law! Merely to have begun upon this enterprise, though the first steps be small and difficult, would be to introduce into all the calculations and judgments of international affairs a new orientation, and into men's lives a compelling purpose.

Albert Einstein, winner of the Nobel Prize in 1921 and perhaps the greatest of all living physicists, started the government's work on the uranium project with his letter to President Roosevelt in the fall of 1939, in which he outlined the possibilities.

Chapter 14

The Way Out

by ALBERT EINSTEIN

THE construction of the atom bomb has brought about the effect that all the people living in cities are threatened, everywhere and constantly, with sudden destruction. There is no doubt that this condition has to be abolished if man is to prove himself worthy, at least to some extent, of the self-chosen name of *homo sapiens.* However, there still exist widely divergent opinions concerning the degree to which traditional social and political forms, historically developed, will have to be sacrificed in order to achieve the desired security.

After the First World War, we were confronted with a paradoxical situation regarding the solution of international conflicts. An international court of justice had been established for a peaceful solution of these conflicts on the basis of international law. Furthermore, a political instrument for securing peace by means of international negotiation in a sort of world parliament had been created in the form of the League of Nations. The nations united in the League had further outlawed as criminal the method of solving conflicts by means of war.

Thus the nations were imbued with an illusion of security that led inevitably to bitter disappointment. For the best court of justice is meaningless unless it is backed by the authority and power to execute its decisions, and exactly the same thing is true of a world parliament. An individual state with sufficient military and economic power can easily resort to violence and voluntarily destroy the entire structure of supranational security built on nothing but words and documents. Moral authority alone is an inadequate means of securing the peace.

The United Nations Organization is now in the process of being tested. It may eventually emerge as the agency of "security without illusion" that we so badly need. But it has not as yet gone beyond the area of moral authority as, in my opinion, it must.

Our situation is rendered more acute by other circumstances, only two of which will be presented here. So long as the individual state, despite its official condemnation of war, has to consider the possibility of engaging in war, it must influence and educate its citizens—and its youth in particular—in such a way that they can easily be converted into efficient soldiers in the event of war. Therefore it is compelled not only to cultivate a technical-military training and type of thinking but also to implant a spirit of national vanity in its people in order to secure their inner readiness for the outbreak of war. Of course, this kind of education counteracts all endeavors to establish moral authority for any supranational security organization.

The danger of war in our time is further heightened by another technical factor. Modern weapons, in particular the atom bomb, have led to a considerable advantage in the means of offense or attack over those of defense. And this could well bring about the result that even responsible statesmen might find themselves compelled to wage a preventive war.

In view of these evident facts there is, in my opinion, only *one* way out.

It is necessary that conditions be established that guarantee the individual state the right to solve its conflicts with other states on a legal basis and under international jurisdiction.

It is necessary that the individual state be prevented from making war by a supranational organization supported by a military power that is exclusively under its control.

Only when these two conditions have been fully met can we have some assurance that we shall not vanish into the atmosphere, dissolved into atoms, one of these days.

From the viewpoint of the political mentality prevailing at present, it may seem illusory, even fantastic, to hope for the realization of such conditions within a period of a few years. Yet their realization cannot wait for a gradual historical development to take its course. For, so long as we do not achieve supranational military security, the above-mentioned factors can always and forcibly lead us into war. Even more than the will for power, the fear of sudden attack will prove to be disastrous for us if we do not openly and decisively meet the problem of depriving national spheres of power of their military strength, turning such power over to a supranational authority.

With due consideration for the difficulties involved in this task, I have no doubt about *one* point. *We shall be able to solve the problem when it will be clearly evident to all that there is no other, no cheaper way out of the present situation.*

Now I feel it my obligation to say something about the individual steps which might lead to a solution of the security problem.

1. Mutual inspection by the leading military powers of methods and installations used for the production of offensive weapons, combined with an interchange of pertinent technical and scientific discoveries, would diminish fear and distrust, at least for the time being. In the breathing spell thus provided we would have to prepare more thorough measures. For this preliminary step should be taken with conscious awareness that the ultimate goal is the denationalization of military power altogether.

This first step is necessary to make any successive moves possible. However, we should be wary of believing that its execution would immediately result in security. There still would remain the possibility of an armament race with regard to a possible future war, and there always exists the temptation to resort once more, by "underground" methods, to the military secret, that is, keeping secret the knowledge about methods and means of and actual preparations for warfare. Real security is tied to the denationalization of military power.

2. This denationalization can be prepared through a steadily increasing interchange of military and scientific-technical personnel among the armies of the different nations. The interchange should follow a carefully elaborated plan, aimed at converting the national armies systematically into a supranational military force. A national army, one might say, is the last place where national feeling may be expected to weaken. Even so, the nationalism can be progressively immunized at a rate proportionate at least to the building of the supranational army; and the whole process can be facilitated by integrating it with the recruiting and training of the latter. The process of interchanging personnel would further lessen the danger of surprise attacks and in itself would lay the psychological foundation for internationalization of military resources.

Simultaneously the strongest military powers could draft the working papers for a supranational security organization and for an arbitration committee, as well as the legal basis for, and the precise stipulation of, obligations, competencies, and restrictions of the latter with respect to the individual nations. They could further decide upon the terms of election for establishing and maintaining these bodies.

When an agreement on these points shall have been reached, a guarantee against wars of worldwide dimensions can be assured.

3. The above-named bodies can now begin to function. The vestiges of national armies can then be either disbanded or placed under the high command of the supranational authority.

4. After the cooperation of the nations of highest military importance has been secured, the attempt should be made to incorporate, if possible, all nations into the supranational organization, provided that it is their voluntary decision to join.

This outline may perhaps create the impression that the presently prevailing military powers are to be assigned too dominant a role. I have tried, however, to present the problem with a view to a sufficiently swift realization that will allow us to avoid difficulties greater than those already inherent in the nature of such a task. It may be simpler, of course, to reach preliminary agreement among the strongest military powers than among *all* nations, big and small, for a body of representatives of all nations is a hopelessly clumsy instrument for the speedy achievement of even preliminary results. Even so, the task confronting us requires of all concerned the utmost sagacity and tolerance, which can be achieved only through awareness of the harsh necessity we have to face.

Chapter 15

American scientists, acutely aware of their responsibilities for making known the full implications of their scientific developments, have been mobilizing ever since the end of the war. In October scientists from the atomic project joined forces, and in December the Federation of American Scientists, open to all scientists and engineers, was formed. Member associations across the country are encouraging qualified scientific and political discussion such as that presented in this book.

Survival is at Stake

by THE FEDERATION OF AMERICAN (ATOMIC) SCIENTISTS

THIS is an unusual book. It has been written by many persons. It is repetitive at times and there is disagreement in its pages, as there is disagreement elsewhere, among scientists and non-scientists, in this country and in other countries. It is a serious and a dangerous circumstance that on this most vital of issues there is not yet the strongest agreement on the basic pattern of the precise solution. The fact that there is not, seven months after the bomb became a reality, tells better than anything else how critical is our problem. For it means that the atomic arms race which can mean our doom is in full swing.

The arms race must be stopped. This book was put together with the sole objective of helping to stop it. This book cannot do what most needs to be done, which is to state the solution. Yet the book has a unity and achieves a purpose even so: it states the problem before us, in full and on sound authority and in one place. Each in his own way, the authors have recognized the nature of the problem and have thereby provided standards by which any proposed solution must be judged.

And this is more than half the battle. There is more to fear from confusion than from disagreement, more from irrelevance than from incompleteness. The many-sided nature of the problem of nuclear energy must be clear from the very content of the book. But just as clear is the common framework within which its authors have written:

¶ The problem has brought us to one of the great crises of history.

¶ The problem has moved onto the political plane and will remain there. Science will devise no defense to make the danger go away.

¶ The problem is a world problem. There can be no merely national solutions.

We of the Federation of American Scientists here undertake to discuss some of the terms of solution, to point some ways to action. But before this we wish to give emphasis to one point which has not been emphasized enough.

If the terror of the bomb is great, and properly great, the hope for man in the release of nuclear energy is even greater. The fruit of that science which follows in the proud tradition of Galileo, the outcome of that complex organization of society which has made possible the city of New York and the Hanford plutonium plant, is the large-scale release of nuclear energy. We cannot now see more than the faint shadow of what such a new force can mean for man. But it is our faith as scientists and our experience as citizens of the twentieth century that it will mean much. It will grow and develop. It will lead a life of its own. No influence that we have seen in our times can prevent this.

Yet it is the eloquent and unanswerable argument of this book that such a growth will bring death to the society that produced it if we do not adapt ourselves to it. This is the dilemma that the release of nuclear energy has brought to a world torn already by a horrible war. *The nations can have atomic energy, and much more. But they cannot have it in a world where war may come.*

There is one way to bring about the needed change, for there is a unique solution. The nations must collaborate for the development of the new force. They cannot, in fact, do otherwise and live. The new energy is, if you like, our common enemy; it must be made our common ally. And it can be done. You have seen how unique are the properties of uranium, how novel the techniques of its control and exploitation. In this fresh field we can proceed better, or at all, the less we are hampered by the old nationalistic conflicts which now divide the world. Too often the controls and safeguards against the misuse of nuclear energy have been discussed as something private and static, apart from its development. Yet it is clear that its development, planned internationally, will simplify and make natural those controls which a haphazard national handling would leave uncertain. Out of the success of such collaboration, moreover, will emerge a greater success. The common possession of atomic energy and the prevention of atomic war will lead us to the end of war itself. That is in the seed of the solution.

What specific properties will the solution exhibit? If we cannot yet outline them in a few pages, we can list for you a few tests by which the genuineness of programs and proposals from whatever source may be assayed.

First of all, our country, the United States, has a peculiar responsibility. We first used the bomb; we alone manufacture it. We are committed no less by the declarations of our leaders than by the existence of the Oak Ridge plants to assume the initiative in devising measures for the control of nuclear energy. *No program is sound unless it recognizes the special duties of the United States, unless it is built upon the principle that our insight and our patience must be greater than that of all the others. The bombs are marked "Made in the U.S.A."*

Second, the year 1946 has a special importance —and the next year, and perhaps the next. Solutions do not grow in a few months, but they must be planted. The time to start is now. The chance for the successful end of the dilemma is greatest while the problem is yet new, while the development of nuclear energy and of the bombs which are its first fruits is still novel and still not widespread. If uncontrolled growth and development are permitted for nuclear energy, we will have lost a unique opportunity. *No program for solution which does not contain steps to be taken at once has recognized the nature of the problem. There is not much time.*

Third, the solution cannot be simply a formal one, although it will certainly bring new rights and new laws. It must be embodied; it must involve an institution, which can spend hard cash and employ earnest and intelligent men. The final form of any agency proposed will be impossible to state, but there must be a beginning. And the initial plan must provide for growth and development in the institution just as the problem will grow and develop. Here, above all, the dead hand of rote must be kept away. Technical and organizational proposals must allow flexibility or they cannot aid us. *Proposals which, on the one hand, imply no material change and require no working staff cannot succeed; proposals which, on the other hand, seek to partition among bureaus the problems of a decade hence cannot succeed either. The problem is a problem of living men and a developing phenomenon. The solution cannot be wholly written on paper.*

There is one more indispensable ingredient but it is not found in this book. It lies in you. The Federation of American Scientists represents men who saw both a hope and a threat in the bomb they worked so long to help create. They have known the facts, they have seen and studied them for years, while still everything was secret. Now the facts are out. They are visible in the rusted rubble of Hiroshima. They are here, in this book, in your hands. Unless these facts become real to you, unless you learn from them as we have learned from them—that we all must act—there will be no answer ever to our problem. Never have people had the opportunity and the responsibility which the citizens of the United States have today. We must learn how to use them, for after an atomic war no good will and intelligence will be needed to bring a permanent peace to the survivors. They will get it in the jumbled stones of their cities.

What can you do?

For one thing, now that you have read this book, discuss it with your friends—don't lay it aside. A great decision rests on how well you and your elected representatives understand and act on the facts and proposals presented in these pages.

Continue your education for survival by being well informed. Ask for the releases and reports on what is happening as prepared by the scientists and issued by the National Committee on Atomic Information.*

Make sure that your Senators and Congressmen know that you are aware of the unprecedented gravity of the problem. Urge them to act with courage and vision in solving the problem of the atomic bomb within the framework of the new ideas that, as this book shows, are necessary to the solution.

Time is short. And survival is at stake.

---

* The address is 1621 K Street, N.W., Washington 6, D. C. Where possible, request material through clubs, unions, or other organizations to which you may belong.